ROGER BANKS

OLD COTTAGE GARDEN
FLOWERS

Illustrated with fifty-four paintings by the author

WORLD'S WORK

KINGSWOOD

For my daughter, Thomasina

Index compiled by Indexing Specialists of Hove

Copyright text and illustrations © 1983 by Roger Banks
Published by
World's Work Ltd
The Windmill Press
Kingswood, Tadworth, Surrey

Filmset, printed and bound in Great Britain by
Hazell Watson & Viney Limited,
Member of the BPCC Group,
Aylesbury, Bucks
ISBN 0 437 01201 8

CONTENTS

Acknowledgements

In writing this book I am particularly indebted to the late Alice M. Coats, pioneer of popular books on the history of plants and often I quote her witty and amusing texts in order to share them with a wider public because, alas! her books were not illustrated.

Many thanks are due to Dr Simpson and the staff of St Andrews University Botanic Garden for allowing me to use their library and paint in their grounds. The Curator and staff of The Royal Botanic Garden, Edinburgh. Cambridge University Botanic Garden, and the British Museum for similar help and privileges.

Among the many people who have helped me with a plant here or a mention of where to find it there, my warmest thanks are due to Mary McMurtrie for help in identification and for the freedom of her garden of old-fashioned flowers at Balbithan, to Donald Lamb, Head Gardener at Balcarres, for his gardening expertise, to Molly Sutton for her expert deciphering of my scrawl into a legible typescript and above all to my wife for her patient forbearance of all this silent scribbling.

' I know a bank where the wild thyme blows
Where oxlips and the nodding violet grows
Quite over-canopied with luscious woodbine,
With sweet musk-roses and with eglantyne,
There sleeps Titania sometime of the night
Lull'd in these flowers with dances and delight.'
A MIDSUMMER NIGHT'S DREAM
William Shakespeare

INTRODUCTION

Everyone knows the familiar combination of Snowdrops and Aconites bursting into flower in the shelter of Winter Jasmine as soon as the waxing sun of the New Year is strong enough to tempt one out of doors on a fine day in late January. The essence of the traditional English garden you might say, but you'd be wrong because Winter Jasmine, *J. nudiflorum*, although it has become deservedly popular, being hardy enough to brighten a cottage porch even in the north of Scotland, was only introduced by Robert Fortune from China as recently as 1844.

To me this information came as a complete surprise. I had taken for granted the idea of Jane Austen's heroines tripping round with Christmas fare through a village sparkling with holly berries and the yellow stars of Winter Jasmine and, as Jane Austen died in 1817, it could never have happened. I began to consider the period aspects of certain flowers just as one would in the design of gardens, houses or furniture. As the eighteenth century gave place to the industrialized nineteenth, and in reaction to it the cult of cottage living gathered strength, what did people grow in their flower garden, that little patch of pure pleasure bounded by the kitchen necessities of fruit and vegetables on one side and, on the other, by the grander world of the big house with its landscaped park, noble trees, orangery and team of professional gardeners?

The closing years of the eighteenth century are just about the last time one can take stock of traditional English garden flowers before the age of scientific plant collecting when the influx of new species from the Americas, Australasias, Africa and Asia poured into Britain to revolutionize our garden flora. Can one visualize spring gardens without Forsythia or shrubberies devoid of Rhododendrons? — no Dahlias and Chrysanthemums? — no herbaceous borders full of multicoloured Lupins and Delphiniums? — no rockeries of Alpines? It challenges one's imagination.

By 1800, the hardier Mediterranean and Near Eastern species had mostly been settled here since Tudor times and gardeners had begun to cultivate new varieties both of them and of our own indigenous wild flowers.

How did plants get here which were neither left behind in the mists by the retreating Ice Age nor imported by later botanists equipped by the nineteenth century with Wardian Cases, travelling glass frames for keeping plants moist and free of salt on long voyages? A species may be introduced on purpose or by accident. Since the great age of exploration by sea opened the New World to us an increasing flow of species considered useful, beautiful or scientifically interesting has poured into the old world from the newly discovered continents. Potatoes and tobacco one thinks of immediately if not, perhaps, of the Scarlet Runner Bean which was brought from South America as a greenhouse exotic, not for its edible pods but on account of its ornamental flowers. Accidental introductions would include the 'African' Marigold which got here via Australia from Mexico and the Boer War in fodder for pack mules or Magellan Ragwort, a large White Waterside Daisy, the seed of which is thought to have come to Caithness from Cape Horn in a cargo of fleeces. Such recent introductions are well documented. Once we get back to the centuries before Linnaeus, classification and nomenclature become progressively more unreliable till we peer questioningly at the dubious woodcuts of early herbals and their vague definitions mixed up with garbled misquotations of classical authors, medieval magic and witchcraft.

'Brought over with the Romans,' school-books always used to say and that was the end of uncertainty to a generation whose history books began tidily with Julius Caesar's landing, neatly organized galleys bringing in vines, Sycamores, Saffron Crocus and vegetables along with proper roads and houses with central heating systems. Now we know too much to accept this simple picture if not yet enough to put others as clear in its place. Botanical history has all sorts of new techniques such as microbiology and carbon-dating at its disposal so that a new age of discovery lies open to us. Whoever previously would have thought it possible to identify from his stomach contents the plant species eaten by the Tolland Man before he was garrotted and pushed into a Danish peat bog in the first century AD or to confirm that the fabric of the Holy Shroud of Turin contains pollen grains of eastern Mediterranean flowers? Such techniques can take us a long way back from the flowers in our cottage garden.

Many of our favourite plants were brought by travellers returning overland to Britain, whether pilgrims, crusaders, traders, ambassadors or those with a taste for botany travelling in the suite of a great man. Some plants were encouraged to grow in cottage gardens for use in dyeing and weaving as well as for medicinal use. Some were brought in to sweeten wine or scent the air in the Middle Ages, and by the Renaissance scholars were already exchanging plants out of botanical curiosity.

Although Egyptian, Arabian and Chinese records are in many cases older, the influence of Classical Greece transmitted to us by Renaissance scholars forms the basis of our botany. Theophrastus, pupil and friend of Aristotle and Plato, and a contemporary of Alexander the Great, wrote our earliest surviving treatise, an *Enquiry into Plants*, in the fourth century BC, and four centuries later his fellow countryman Dioscorides wrote *De Materia Medica*, on the use of herbs, which became the foundation of medicine for medieval apothecaries. It is hard to realize that so many of the flowers we regard as part of the character of an old English garden were originally introduced for quite other purposes, such as Christmas Roses for the purgative value of their black roots, Crocus for saffron and Houseleeks for healing. Only with the Renaissance of classical learning and the sudden outburst of interest in all subjects, the New Learning, do we recognize botanists and gardeners as sharing our interest in plants for their own sakes.

In Tudor England we have William Turner, a friend of Ridley and Latimer, who complained that whilst at Cambridge in the 1530s he 'could learne never one Greke nether Latin nor English name even amongst the phisicians of any herb or tre, such was this ignorance in simples at that tyme'. He took his degree in medicine at Bologna University and collected plants in the Rhineland, probably introducing Lucerne as a fodder crop to Britain. Though his life was largely taken up with theological controversy, William Turner was the first Englishman to study plants scientifically, thus earning his soubriquet as the Father of English Botany. He published his *Newe Herball* in 1551. A generation later Henry Lyte, eleventh lord of the manor, in direct line, of Lytescary in Somerset, had translated Rembert Dodoens' *Cruydeboeck*, first produced in Antwerp in 1554, and published in 1578 *A Niewe Herball* or *Historie of Plantes*, dedicated with commendatory verses to Queen Elizabeth. Gerard's famous *Herball*, or *Historie of Plants*, published in 1597, is now largely discredited as being considerably filched without acknowledgement from Dodoens, but its chief interest for us lies in Gerard's notes of the whereabouts of scarce plants in England. He was a skilled herbalist and superintendent of Lord Burghley's garden in the Strand and his list of plants of his own garden in nearby Holborn is the first catalogue of any one garden.

In the next century in a continuing tradition we have John Tradescant, who is described in an Ashmolean manuscript now in the Bodleian as a 'Wustersher' man and an 'industrious searcher and lover of all nature's varieties'. In 1617 he 'paid 25L for the transport of one person to Virginia'. It was not himself, though next year he went on 'A voiag of ambassad undertaken by the Right honnorabl Sr Dudlie Digges in the year 1618' to Archangel, giving us our earliest account of plants of Russia. In 1620 Tradescant joined as a gentleman volunteer against the Algerine corsairs and brought back apricots. In 1625, in the service of the Duke of Buckingham, he wrote that his job was 'to deal with all merchants from all places but especially from, Virginia . . . for all manner of rare beasts fowls, shells and stones.' After Buckingham's death Tradescant entered the service of King Charles and his Queen Henrietta Maria. Tradescant's Ark, established in Lambeth as a repository for his collection of natural curiosities, was one of the earliest, and the building housing it was pulled down only in 1881. His son, also John, was in Virginia in 1637, 'gathering all varieties of flower, plants, shells etc.,' for the collection at Lambeth. He succeeded his father as gardener to the Queen and introduced Lilac, Acacia and Plane Tree. Elias Ashmole, who had assisted in the cataloguing of the collection of Tradescant's Ark, was bequeathed in Tradescant's will 'my Closet of Rarities', including a stuffed dodo and a 'Phoenix' Tayle', which in turn became eventually the nucleus of the Ashmolean Collection in Oxford University.

John Parkinson, 1567 to 1650, herbalist and apothecary of Long Acre, also had a garden 'well stored with rarities'. On the publication of his *Paradisus Terrestris* in 1629 he obtained from Charles I the title of Botanicus Regius Primarius. The work contains thirty-three first mentions of native species and seven new introductions. He later became apothecary to James I. Parkinson cherished, as he put it, 'a garden of all sorts of pleasant flowers which our English ayre will permit to be nursed up.' In 1640 his second great work was published, *Theatrum Botanicum, The Theater of Plantes or An Universall and Complete Herball*. It mentioned 3,800 plants, surpassing Johnson's edition of Gerard, revised in 1633 still with only 2,850 entries, so that most of our old cottage garden flowers are mentioned in it. In 1884 a Parkinson Society was founded by a Mrs Ewing 'to search out and cultivate old garden flowers, to plant waste spaces with hardy flowers, and to prevent extermination.' Now it is sadly defunct.

In Parkinson's own seventeenth century, interest in horticulture grew ever more purposeful. Flemish and Dutch merchants were prepared to pay for consignments of bulbs from the Near East and plant breeders, known as Florists, soon got to work on them. Tulipomania and the enthusiasm of certain groups of Florists for particular species, such as Auriculas, Polyanthus, Pinks and Pansies, showed how quickly spectacular developments can take place but by 1800, before greenhouses were invented, only the hardiest of easily grown flowers could be described as 'cottage garden'.

My aim as a painter is to give some idea of the popular species which were generally available to the ordinary gardener without specialist knowledge or friends in high places at Kew; flowers which in many cases have almost disappeared and, their days of fashionable cultivation over, linger unappreciated in a few half forgotten country gardens.

I realize that such an aim raises as many problems as it would solve. How can one draw a line between the flowers of cottage gardens and those of grand establishments? At either end of the scale the difference between carefully nurtured exotics and homely common flowers is obvious but what about those in the middle ground? Only a suspicion of my publisher's rising blood pressure as a dozen extra plates grew to include still more forced me to grab the pruning knife. Fair Maids of France are in because they have fallen from their erstwhile popularity. Crown Anemones and Oriental Poppies are out because their strident reds wouldn't easily combine with more gentle flowers on a plate. Where I have had to choose between them, a well known herbaceous plant popular today and illustrated in every garden book has given place to a Tudor introduction not in fashion and only surviving in a few old gardens. Whereas peripheral species such as the Childling Pink are easily slipped in because of their size, for the same reason grand and important plants such as Wake Robins and Bears' Breeks, each needing a plate to themselves, have regretfully been omitted. In mitigation may I point out that I have dealt with the edible ones such as Sunflowers, Elecampane and culinary herbs in *Living in a Wild Garden*. Our handful of indigenous evergreen trees are in because they are so few and with hedgerow species important as part of the traditional frame for and ecology of the cottage garden.

English names are used where possible but the Latin name is always included too because, as I hope I shall show, the most amusing side of the history of plants is confusion over what was what in the world before the advent of the great Swede Gustave von Linné. 'L' coming after the Latin signifies that the scientific name was fixed by Linnaeus, as he is known, the father of botany and organizer of our system of nomenclature. The Oxford Dictionary explains it:

> Under the family or sub-family the generic and specific names together form the scientific proper name of an animal or plant, the generic name standing first, and being written with an initial capital.

The basic idea is as simple as a telephone directory; for John Brown see 'Brown, John' and under 'Smith', Mary or Ebenezer Smith. Difficulties come, as with the telephone directory, when names change by marriage or divorce. A botanist working on the family Smith discovers so many relations of Ebenezer's that are so unlike Mary as to constitute a genus on their own, the Ebenezers, or 'Brown, John' is discovered to be exactly the same species as one identified and named in China a hundred years previously. One of them has to give way and so loses his separate identity. *Eranthis hyemalis* (L.) Salisbury means that at sometime in the centuries since Linnaeus there has been some confusion or change of name but that Salisbury has reclassified the genus and confirmed Linnaeus's original nomenclature. Scientific rectitude, though it may appear to be unnecessarily pedantic in a popular book, is particularly important when dealing

with old-fashioned flowers because the same plant may appear through the centuries and in various places under half a dozen different names.

As the year begins, the shortest days of winter are soon brightened by the yellow stars of Jasmine. The shrub straggles in untidy thickets against a wall but the seemingly dead growth supports a waterfall of blossom and makes perfect winter shelter for wrens. Because they are insectivorous and live their lives close to the ground in gardens where there is a sufficiency of low cover, they are one of the most important birds. As Wren inspects the possibilities of food supplies and nesting sites under Winter Jasmine, its downward draining twigs guarantee a warm spot for bringing the first bulbs into flower.

Aconites are the flowers of Hecate, Queen of the Underworld, and aconite was the classical poison for which there was no antidote. Related to Monkshood and sometimes together with it called Wolfsbane, the plant had practical use as a bait for predators marauding round the sheepfolds in winter.

According to Henry Lyte writing in 1578, 'These venomous and naughtie hearbs are found in this countrie planted in the gardens of certain Herborists'. So they must have been of earlier introduction though, with their little ruff of green leaves, they were long confused with Ranunculus and Hellebores. *Eranthis*, flower of the earth, and *hyemalis*, of winter, was the Latin name bestowed on them in 1807.

Both Snowdrops, *Galanthus nivalis*, L. and Aconites, *Eranthis hyemalis*, (L.) Salisb., are a familiar sight growing wild in sheets of white and yellow under the bare branches of coppice woodland. Though their introduction is too long ago for an exact record, they appear with an asterisk in books on wild flowers signifying that they are not admitted to be part of our indigenous flora.

Galanthus, Milkflower, was we learn a name given by Linnaeus to distinguish the flower from *Leucojum*, Snowflakes. 'Bulbous Violet' was the name used by Gerard in his *Herball* of 1597 and by Parkinson in his *Paradisi in sole Paradisus terrestris* of 1629. They are mentioned by neither Chaucer nor Shakespeare though, as Alice Coats, pioneer and historian, remarks with refreshing commonsense, 'how could a poet be expected to refer to a flower which was known, if it was known at all, by the name of "bulbous Violet"?' She points out that 'Not until Johnson brought out the revised edition of Gerard's herbal in 1633 do we read: "Some call them Snowdrops".' Crispin de Pass, writing in the Netherlands in 1614, says 'This plant abounds in Italy, but is not found here, except in the gardens of the curious!' Thus our homely Snowdrop spreading wild in the woods

Winter Jasmine sheltering early Snowdrops and Aconites. Spring Snowflakes and a Wren prospecting nesting sites.

is, it seems, probably imported comparatively recently, botanically speaking, from Italy. The Spring Snowflake *Leucojum vernum* L., on the other hand, may have originated in the damp woodlands of Wessex. With its stronger, green, strap-like leaves and larger creamy bells, delicately cusped and marked with green, two or three to a stalk, it has become a cottage garden favourite. Its close relative, *Leucojum aestivum* L., the Summer Snowflake or Loddon Lily, is slightly smaller and blooms sometime later. It is a definite native still surviving on a few islands in the Thames and therefore is the object of an annual waterborne pilgrimage by the Botanical Society of the British Isles (BSBI) to make sure that all is well. So we learn that the rarity is a native wildflower and the common or garden Snowdrop an introduced species. In fact, the *Horticultural Journal* mentions a dozen different varieties of Snowdrop.

In addition, scientists have now shown us that just as the concave star of an open Crocus is designed to reflect the sun's rays to concentrate maximum heat at the stigmas as an encouragement for pollinating insects, so the inverted bell of a Snowdrop works as a heat trap, conserving the day's warmth to a difference of as much as two degrees in the cold nights of February. As we begin to understand flowers as highly efficient working parts in conjunction with their pollinators in the process of getting the plant's seeds fertilized, simple flowers of our cottage gardens are seen to have as exciting private lives as any exotics in David Attenborough's jungle.

As I look at Snowdrops collapsed on the lawn after a night of frost, I can see how the 'doubles', expensive fancies from a bulb merchant's catalogue, are less efficient at raising their heavier infertile heads than the simple 'singles', which perk up and dance in the warming sun, waiting for the bees. It is that sort of thing which leads me to begin to understand my garden as a living organism rather than just a lot of meaningless colour. Because I want to know about a flower's pollinators and something of its growing conditions, whether it prospers in a changing mini-climate even within the confines of a cottage garden or the deep leafmould of woodland shade or has survived the centuries by colonizing the few feet of desert on an old wall-head, I hope I shall not be taken to task for being insufficiently botanical' if I extend a little into this natural ecology of flowers beyond what has hitherto been considered conventional.

SPRING

Crocus, because of the highly-prized spice obtained from its dried stigmas is one of the oldest flower names in the world. From Latin 'Crocum' it goes back via 'Kurkuma' in Sanskrit and 'Karkom' in the Song of Solomon to the basic 'KRK' in the consonantal root tongues of the ancient East. In Arabic the spice is 'Za-feran' which we anglicize into saffron.

The Romans are thought to have brought their wild *Crocus sativus* to Britain from the Mediterranean and returning crusaders to have reintroduced it in the Middle Ages. Concentrated round Saffron Walden in Essex and selling at thirty shillings a pound in the eighteenth century, saffron was a valuable crop producing up to five pounds an acre depending on plentiful labour to pluck the tiny stigmas from over four thousand Crocus flowers to obtain an ounce. It is now grown only in Spain but Miller's *Gardener's Dictionary* of 1768 contains a good account of the industry.

Crocus sativus, with its large drooping stigmas, is an autumn flowering species and Linnaeus, living in northern lands and never having seen a Crocus, made the mistake of thinking that all other spring Crocuses were varieties of it. On the contrary, the early flowering *Crocus flavus*, formerly *Crocus aureus* or Dutch Yellow Crocus, may be recognized in Gerard's *Herball* published in 1597. His description of it 'with flowers of a most perfect shining yellow colour seeming afar off to be a hot glowing cole of fire' is as vividly recognizable today as four hundred years ago. Gerard received it from his friend Jean Robin, the King's gardener, in Paris and it was first recorded by Charles de l'Ecluse, tutor to the rich banking Fugger family, who had it sent to him in 1579 from Serbia, the most westerly extent in the range of this Balkan and Turkish species. Under the Latin version of his name, Carolus Clusius, de l'Ecluse had published a first flora of Iberia in 1576 and included a woodcut of *Crocus aureus* in his *Rariorum Plantarum Historia*, Antwerp 1601.

Crocus vernus is the tiny, white starred species which comes forth in drifts from Alpine turf as the snows melt under the returning sun. Although it can't adjust if transferred directly from its spartan home to the mild lowlands, it is a very variable species and all our purple, white and striped garden varieties such as 'Madame Mina' are ultimately derived from it and the Imperati group of Ligurian coasts, which have biscuit-coloured buds opening to clear violet interiors.

C. tommasinianus, closely related and hybridizing with equal alacrity, was brought back from the Dalmatian coast by Dean Herbert in the early years of Victoria's reign and named after his botanist friend, Signor Muzio di Tommasini of Trieste.

Mezereon tempts out the first Honey-bee. In moss beneath the Crocus a Bumblebee has her nest.

E. A. Bowles, in his authoritative *Handbook of Crocus & Colchicums*, distinguishes the two species by illuminating botanical jargon surrounding the two groups with an apt simile:

'*C. tommasinianus* has a slender flower bud due to the close wrapping of the segments, reminding one of a smart new umbrella compared with the thicker clumsier buds of *C. vernus*, which are more of the "gamp" style.'

The Scotch Crocus is an old garden form of the Balkan blue and white *C. biflorus* distinguished from the Saffron Crocus by Miller in the *Gardener's Dictionary* of 1768 as 'with a few narrow leaves which are closely wrapped round by a spatha or sheath, out of which rise two flowers'. Like other old garden varieties it is sterile and increases rapidly by corm division. Bowles considers it 'likely to have originated as a seedling in some Scottish garden and was introduced from thence to England and Holland. Nothing very closely resembling it has been found in the wild state'. With *C. aëreus* and *C. chrysanthus* it is out of the annulate group in which the lower part of the corm splits into over-lapping rings and Maw, in the nineteenth century, said the former hybridize naturally on the Bythinian Olympus as well as with *C. biflorus* in gardens, so that seems our best clue to origins of this charming flower and parent of our modern 'Blue Pearl' and 'Cream Beauty'.

Mezereon, *Daphne mezereum* L., is another name deriving from Arabic: 'destroyer of life' because the whole plant is poisonous to men, the red berries extremely so. The genus name 'Daphne' is founded on mistaken identity because the plant has only superficial resemblance to Bay Laurel, *Laurus nobilis*, with its associated legend of the nymph Daphne, pursued by Apollo, being turned into a bush to escape rape.

Some have doubted that it is a British native because it was only first recorded wild as recently as 1759. As *D. mezereum* blooms in February and in the wild tends to creep through sparse underscrub, its branches weighed down by drifts of dead leaves to layer naturally, this is understandable. Today, though it is still found wild in a few calcareous woods from Hampshire to the Pennines, it is in the garden where it grows into a neat little bush that we know it. Though hardy, Mezereon is temperamental, a healthy bush suddenly dying for no apparent reason. This trait doesn't matter so much when its red berries will be found to have produced a new generation of seedlings in the shelter of their dead parent, but Alice Coats cites a fascinating report on the propensity of greenfinches in urban districts for learning to eat the berries while still green which may inhibit the supply of seedlings. For those who find its pink too sticky-sweet in colour, *D. mezereum* also has a rarer white form.

Daphne laureola L., sometimes called Spurge Laurel, though it is no more a Spurge than it is a Laurel, is a recognizable relation of Mezereon. It is often present in limestone woods south-east of a line from Dorset to Yorkshire for years without being noticed, so demurely reticent is it with its scented waxy flowers, greenish yellow and hidden in clusters under a palm-like spread of leaves at the top of its woody stalk. This lack of show is possible because *D. Laureola* is pollinated by moths, an uncertain arrangement one would have thought for a species flowering so early in spring. Night-scented flowers combined with its neat evergreen growth made *D. laureola* popular as a house plant in the eighteenth century. According to Miller 'There are poor People who get the young plantes out of the Woods, and carry them about the Town to sell in Winter and Spring.' Under the names Lauriel, Lovel and Lowry, it was well known as a plentiful hedge plant to Turner as far back as in 1548, whereas *D. mezereum*, though cultivated from Polish stock introduced by Gerard and known in sixteenth century gardens, had to wait two more centuries to be found here in the wild.

Appreciating similar limestone woodland habitats across Euro-Asia, Violets were used as the basis of a sherbet drink by the Persians and for making chaplets by the Greeks long before being candied as a sweetmeat in this country. John Evelyn ate young leaves as well, fried and sprinkled with lemon and sugar.

Botanically speaking, Wild Violets are classified into two groups, those whose leaves enlarge after flowering and those whose leaves do not, which may seem odd to those who dream of Violets in a nest of fresh green leaves in spring without inquiring further. This would be a pity because most people want Violets for their scent and most wild Violets scarcely smell at all. The Common Violet, *Viola riviniana*, belongs to the second group with hardly any smell but because it has a tidy growth is encouraged along the edges of most cottage garden paths. The Dog Violet, *V. canina*, is also a Violet which branches neatly from the base and is common on sandy heathland. Its name is nothing to do with dogs but, as in Dog Rose and Dog's Mercury, alludes to the season when the Dog Star, Sirius, the brightest of the fixed stars in the constellation of the Greater Dog, is in the ascendant.

V. odorata L., the Sweet Violet, is however, a plant with stolons: it runs about untidily like *V. palustris*, the Bog Violet, and during the summer produces coarse old leaves which live on through the following winter to protect the young shoots next spring. To have Sweet Violets for their scent one has to put up with their untidy habit of growth. On a

warm bank in the shelter of a cottage garden hedgerow, this is no problem. *V. odorata* has now been recognized as having, in addition to the violet blue type, three subspecies: var. *dumetorum* (Jord.) Rouy and Fouc., flowers white with a red spur, var. *praecox* Gregory, with flowers small and darkly purple, and var. *subcarnea* (Jord.) Parl. with pink flowers.

Today, oil from Sweet Violets is distilled for toilet water and the flowers are crystallized in gum arabic and rose water to put on cakes and chocolates but the flower has been known since the ancient civilizations of the East and used to alleviate sleeplessness.

Sweet Violets have been cultivated in Britain since the Middle Ages especially in the milder climate of the South-West, and became great favourites of the Victorians even if nowadays their popularity has waned. The runners may be pegged down like strawberries to produce plenty of new plants and in the garden a good mulch reproduces the annual layer of dead leaves in the old fashioned hedgerow.

Roy Genders, the authority on cultivating old fashioned flowers, says that many of the old favourites are susceptible to red spider and mildew and that is why commercial growers favour varieties resistant to pest and disease even though their large handsome blooms are devoid of scent.

Dog's Tooth Violet, *Erythronium dens-canis* L., is neither a Violet nor recognizably like a dog's tooth till you dig it up and inspect its roots, all right with a dried specimen in a herbarium but not a practice calculated to keep your friendship with a species so notoriously intolerant of disturbance. To the Tudors, 'violet' was a vague term applied to anything from a Snowdrop to a Lupin which makes even the most reluctant Latin scholar see the virtues of Linnaean classification in which *Violaceae* can be shown to be a term of scientific accuracy denoting solitary flowers with five unequal petals, the lower ones being spurred, which includes Pansies and Violets but not Erythroniums. One can understand the confusion because, like many Violets, the pink Erythronium is a flower of limestone turf, common enough on the Alpine pastures of Central Europe. It was probably first grown here in Elizabeth's reign by de l'Obel, after the Latin form of whose name Lobelia is named, and who became superintendent of Lord Zouch's physic garden at Hackney and later botanist to James I. His contemporary Gerard tells us why Erythronium continued in such esteem for the next hundred years or so. Because of its spotted leaves it was 'supposed to be the Satyrion of Dioscorides', a classical aphrodisiac. I should like to find glowing testimony of this attribute but even such a redoubtable scholar as Alice Coats can unearth nothing more exciting than an eighteenth-century professor of botany in St Petersburg, one Gmelin, who said that Tartars made 'nourishing and excellent' soup of it.

Above a bank of Sweet Violets, Spurge Laurel has a moth pollinator.

In 1665 the first of the American Erythroniums began to be grown here, mostly flowers in shades of cream but with leaves similarly spotted and blotched, and we hear no more of the Satyrion. Perhaps the learned climate of the Royal Society founded by Charles II was as above such superstitious aids as the royal bed was needless of them.

Nowadays we can derive a more innocent thrill waiting for the spotted leaves of Erythronium, flickering and almost reptilian as they push through the turf in early spring. As for difficulty in moving them, I find that provided they are grown naturally in damp mossy turf, large clumps may be divided into pieces after flowering and if buried in plenty of old mortar to remind them of their native Dolomites and plenty of leafmould to prevent drying out in summer, they soon settle down to flowering again in their new positions.

Anemone pulsatilla or, as we now call it, *Pulsatilla vulgaris* Mill., with its purple trumpet flowers nestling between and contrasting with the soft furry grey of its leaves and pennant seedheads, has been cultivated as a garden flower since the sixteenth century but is in its origin a British native of chalk downland. Near Oxford on the Berkshire Downs and south of Cambridge on the Gogmagog Hills it was once common, too common to appeal to the medieval apothecaries who had no medicinal use for it, beyond obtaining a bright green dye as recorded in the household accounts of Edward I for the dyeing and gilding of four hundred easter eggs at Court.

Pulsatilla's striking appearance and natural habitat in downland turf along the prehistoric Ridgeway made it a cult flower from earliest times. Christians adapted it for Easter as the Pasque Flower. Later on, people said that it would grow only where Danish blood – or Saxon, or Ancient British, according to one's romantic historical allegiance – had been shed, because Pulsatilla likes just those same strategic sites as army commanders chose for making their last stands in defensive forts. Such legends led a past generation of Florists to dose their poor Pulsatillas with the floor-sweepings of an abbattoir when it really wants the opposite treatment, to be left alone in the dry calcareous turf favoured by Cowslips. It does well in cottage gardens because there is often lime washing off the house, crumbling mortar at the side of a path, good drainage occasioned by some steps or pavement providing a little corner for it to thrive in. Though these garden-grown flowers are larger than the wildlings, they are the self-same species as those growing wild on the downs since prehistoric times. As in so many pink or purple species, such as Heather, a white form is sometimes produced naturally. A glance at the Botanical Atlas shows that prior to 1930 Pulsatilla was found at many sites up the eastern, drier side of the Pennines as far north as Teesdale, whereas now it is restricted to the Cotswolds, the Ridgeway and a few surviving places on limestone in the East Midlands.

As there can be no reason for this British wildflower's not being common still, except man's rapaciousness, I suggest that it would be a suitable species for us to begin making amends for our past collective vandalism by replacing it in suitable habitats of exposed limestone outcrop such as in motorway cuttings. This would not disturb the distribution of species in existing reserves but act as pump-priming operation so that Pulsatilla and similar wildflowers would have an opportunity to recolonize areas where they were once common. Motorways, because of their inaccessibility to people in the 1980s, give us that chance. In an ideal world this would be done by County Naturalists Trusts but for those individualists who can't wait for Utopia, try making little mud bombs studded with seeds and throwing them out of the roof as you cruise round those huge land-wasting roundabouts!

Cowslip, *Primula veris* L., is a borderline species. As a wildflower of dry calcareous turf it was formerly found and should still be taken for granted in any suitable locality whether on chalk downland, limestone uplands or where broken shells make up the requisite lime content in the turf of seaside links. Unfortunately it is now less common than it was due to the usual reasons: improving farming and ploughing up of old pastures, acidic aerial pollution of which *Primulaceae* are especially intolerant, being over picked and dug up. Although this last is now illegal, it seems hard to stop a child picking Cowslips when years ago they all knew how to make 'tossyballs' of them in spring, just as their elders made cowslip wine. The trouble comes when motorized urban millions descend like locusts on our dwindling stock of cowslip meadows. It's all a matter of balance and an instance where a new informed type of cottage gardener can do a lot to help. When altering an old house, don't allow the builder to take away the rubble. There's always a sunny corner where drainage can be improved by a raised bed of bricks filled with rubbish. Plastic, cans and broken glass may all be buried safely because it is never going to be disturbed. Top it off with leafmould, sand, old mortar and turf as a bank for Cowslips and similar calcicoles, as flowers appreciative of chalk are known to scientists. Hand weed your bank till the turf is established sufficiently to keep out nettles, cow parsley, ground elder and similar rank growth. Then scythe it annually. Cowslips may be grown from seed obtained commercially and transplanted when large enough to fend for themselves. Cranesbills, Columbines and Pulsatillas, as they are typical wildflowers of calcareous turf, also respond well to such treatment.

Primrose, according to Roy Genders, 'takes its name from primaverola, a diminutive of *prima vera* meaning the first flower of springtime.' It developed into 'primerole', and one of the earliest writings to mention this name was that of Walter de Biblesworth in the late thirteenth century:

Primerole et prime veyre

Sur tere aperunt Entems deveyre.

The name soon became 'primrose' which, in Shakespeare's time, was a plant held in such great esteem as to be the word most often used to denote excellence, as when Spenser wrote 'She is the pride and primrose of the rest'.

Primula veris is the darker yellow Cowslip and the paler yellow Primrose is *P. vulgaris*. By Tudor times, variations were well known. Tabernaemontanus described the Double Yellow Primrose in 1500 and the Double White Primrose (var. *alba plena*) was described by Gerard in his herbal. When Primroses and Cowslips hybridize the result is properly described

In downland turf grow Cowslips, Pasque Flowers (including a white form), and Dog's Tooth Violets.

as *P. veris x vulgaris* or False Oxlip. The true Oxlip, or Paigle, *P. elatior* (L.) Hill, has the spade-shaped leaves of a Cowslip and the pale yellow flowers of a Primrose but the bells hang to one side and their throat is more open than the hybrid's. The Oxlip only grows wild on boulder clay round Cambridge where, as it is liable to hybridize with any of its cousins, confusion is worse confounded.

The Hose-in-Hose takes its name from Tudor stockings which were usually put on two pairs at a time — an interesting comment on their quality. The lower bloom is really a petal-like calyx and in some forms this may be striped green, red or yellow, giving rise to the Pantaloon Primrose.

The Jack-in-the-Green Primrose, later known as Jack in the Pulpit, has a ruff of leaf-like bracts backing the flower which may be yellow or even pink. Sometimes this aberration is even more wild and the result is then a Franticke or Foolish Primrose, described in 1629 by Parkinson: 'It is called "Foolish" because it beareth at the top of the stalk a tuft of small, long, green leaves with some yellow leaves, as it were pieces of flowers broken and standing amongst the green leaves.'

Such 'anomalous inflorescences', as they are known scientifically, still turn up even in the stock of commercial growers, witness the exhibit shown to the Botanical Society in 1981 which showed what appeared to be 'Jack-in-the-Greens' well on the way to being 'Franticke' all among thousands of otherwise unremarkable primroses destined for the flower trade.

Pink Primroses are another ancient hybrid probably deriving their colour from some forgotten Polyanthus and in turn descended from the Turkish purple but often surviving the centuries in cottage gardens.

Polyanthus, 'many flowered', comes from Greek and the flower was first described in 1665 by Rea in his *Complete Florilege*: 'The red cowslip or oxslip is of several sorts, all bearing many flowers on one stalk . . . some bigger like oxlips.'

The flower is believed to derive from the False Oxlip, *P. veris x vulgaris*, crossed again with the Red or Turkie-purple Primrose brought back by John Tradescant from the Black Sea coast. Its colour was always reddish purple until, only a hundred years ago, the great garden designer Gertrude Jekyll discovered a yellow one in her garden.

In the latter part of the eighteenth century, however, we know from John Hill's *The Vegetable Kingdom* of 1757 that Polyanthus with gold lacing to the edges of their dark red petals had begun to appear. Their parentage is unknown but Alpine Primulas may have had a part in it. Certainly this striking feature had an immediate appeal for flower lovers.

Elizabethan Primroses include (reading clockwise) Jack-in-the-Green, Turkey Red, Common, Double White, Hose in Hose.

Roger
Banks 4/8?

In 1759 James Justice wrote from Dalkeith that 'the varieties which are obtained every year by the florists who save and sow their seeds, are very great' and in 1769 we have the first reference to an exhibition of the plants at the Lichfield home of John Barnes.

For a century the cult ran its course, producing, at its peak in 1860, several hundred different varieties of both gold and silver-laced Polyanthus, nearly all of them since lost and long forgotten.

These Florists were usually artisans of the early industrial revolution, intelligent craftsmen capable of taking immense pains in their tiny cottage gardens or backyards, getting the potting soil and conditions just right for producing their particular fancy.

Primulaceae in general are intolerant of pollution because of their deeply furrowed leaves which they have evolved for trapping every raindrop into the centre of the plant. This perfect irrigation system counts against them in places where there is industrial grime or, nowadays, on roadside verges with too many exhaust fumes. As the nineteenth-century towns produced widening circles of sooty grime, so larger mills and factories absorbed the once independent townsfolk and the growing of laced Polyanthus became an almost forgotten chapter in the history of cottage garden flowers.

The conditions provided by light woodland with deep leafmould never drying out in summer, sheltered in winter, yet open to the warmth of the spring sun are perfectly suited to Polyanthus, as Vita Sackville-West realized in our own century when she established the gardens at Sissinghurst in Kent where the Primrose family and Hazelwood go naturally together. Hazel coppice was part of rural industry, the poles being needed to make hurdles, fruit baskets and supports for hops and beans while bundles of twigs were used as peasticks. Cobnuts from Hazels in the cottage garden hedge were an extra bonus when the days drew in before Christmas.

Primula auricula, commonly called Auriculas or Dusty Millers, came from the Alps. Our native pink Birdseye Primrose, *P. farinosa*, L., is restricted to Pennine limestone and the Scots Primrose, *P. scotica* L., is shyer still and only to be found on the northern coasts of Sutherland, Caithness and Orkney. The Scots Primrose is the exemplar of everything a garden plant can never be however hard horticulturists strive; brilliant magenta flowers with primrose eyes held in a crisp rosette of lettuce-green leaves yet the whole plant, so delicate that jewellery is the only possible analogy, will fit into a small coffee cup. Its habitat is spume-swept turf, and it is proportionately tough. Though easy to transplant it never survives for long away from home because *P. scotica* will only grow where

Gold and silver laced Polyanthus with a natural
hybrid Primrose and Hazel Catkins.

the Glaucous Sedge, *Carex flacca*, and the Sea Plantain, *Plantago maritima*, abound: find the two species together and the third will be there if you look. So few will take the trouble to go and see: botanical illustrators are as bad as gardeners and *P. scotica* is an example of the traps that lie in wait for wildflower painters loath to leave the comfort of their studios to climb mountains or descend into bogs to see how their subjects really grow. Even the Scottish Wildlife Trust, whose emblem it is, got it wrong on their publications cover. No one unfamiliar with *P. scotica* in situ can believe that it is so brilliant yet so tiny but, of course, even an extra inch of stalk would be fatal for a plant regularly swept by sub-Arctic storms and unable to set seed below two inches, sheep-grazing height, on the links where it lives.

Although the Birdseye Primrose has now found favour in modern rock and bog gardens, the old Mountain Cowslip, sometimes called Bears' Ears from the shape of its leaves and known to Latin Scholars as *Auricula ursi*, was brought to this country in the sixteenth century probably by Protestant refugee weavers. Gerard published a description of it in his *Herball* of 1597, stating that they grow naturally 'upon the Alpish and Helvetian mountains' and that 'most of them do grow in our London gardens'. Clusius, then court botanist to the Emperor Maximilian II and a pioneer in acclimatizing Alpine plants to the lowland gardens near Vienna, knew that Auriculas were found wild in the mountains above Innsbruck. He grew two types, yellow and chocolate with a yellow eye, which became known as *Auricula ursi* I and *Auricula ursi* II, and there the scientific matter of origins rested for nearly three centuries.

Parkinson, writing in 1629, realized that Auriculas' foliage belonged to two main groups, those with white mealy leaves that we know as 'Borders' and those without any trace of mealiness that we call 'Alpines', but due to the elementary botany of the day, he described twenty sorts including some which, we now realize, were probably not even Primulas.

Thereafter fanciers took over developing a variety of colours in different combinations. According to John Evelyn's friend, Sir Thomas Hanmer, writing in his garden book of 1659, 'We have whites, yellows of all sorts, haire colours, orenges, cherry colours, crimson and other reds, violets, purples, murreys, tawneys, olives, cinnamon colours, ash color, dunns and what not?'

As the Auricula is very hardy and free of insect pests, it proved an ideal plant for amateurs to cultivate in seemingly endless variety. Striped sorts were developed and were produced in the eighteenth century, some commanding up to twenty pounds a plant. Paste and edging of the petals in various colours with or without meal proved a delight to Florists and

Old varieties of Auriculas or Dusty Millers and a Bumblebee.

judges' rules show the lengths to which they went:

'Growers of the Show Auricula long since set up standards of excellence in their seedlings and at the present time the following points are regarded as essential. The truss should consist of an odd number of flowers or 'pips'. It should be carried erect on a stalk rising from the centre of the foliage long enough to lift it well above the leaves but not looking drawn, the stalk and the pedicels should be proportionate to the foliage and the truss. The flowers themselves must conform to a very strict standard. They must be flat and smooth and should have six segments. The mouth of the tube should show the stamens. If the pistil shows it is a defect. The tube which should measure $^1/_{16}$ in. across, must be yellow. The tube is surrounded by the eye which must be a regular ring covered with white farina measuring $^9/_{16}$ in. from edge to edge. Surrounding this is the body colour, intensely black or at least very dark, free of farina, with rays evenly passing a short distance into the border; this ring is half the width of the paste. Outside this is the edge which determines the type of Show Auricula. The whole flower measures $1^3/_{16}$ across. There are 5 types . . .'

Phew! It's hard to remember that it's a living flower they're talking about.

Among famous varieties, white edged 'Hortaine', and green-edged 'Rule Arbiter' were known to commercial growers as early as 1757 and many green and white combinations were developed from them. As for the green Auricula of Regency prints, the geneticist and plant breeder, Sir Rowland Biffen, writing on the Auricula in 1949, points out that 'this owes its colour to a superimposed layer of lavender blue on a yellow ground. Its colour has nothing to do with that of the green edges.' On the other hand the green edged Auriculas, so popular at the end of the eighteenth century, are now recognized to have thrown a green sport, in cell structure the same as a leaf; freaks of the same order as Jack-in-the-Green Primroses and the Green Rose. Even nowadays very little is known of this scientific phenomenon known as 'virescent mutation'. However, once recognized, Anton Kerner, an Austrian botanist working at the end of the nineteenth century, was able to show that not only was the rare rust-coloured species described by Jaquin as *Primula pubescens* a natural hybrid between the scented yellow *P. auricula* of limestone scree and pink *P. hirsuta* of granite schist but that, as it was unusually fertile, all our garden Auriculas are descended from it, the whole range of their colours lying 'within that anticipated from a knowledge of the properties of hirsutin (the pigment of *P. hirsuta*) and the pigment of *P. auricula*'.

Today, experts divide Auriculas into four groups; show, alpine, border and later miniatures. The hardy easily grown border types are all that concern us in the cottage garden.

Like other species, Alpine in origin, they still stand any amount of cold but not puddles of water around their roots. When I was a child my mother grew Dusty Millers on cushions of leafmould about the clay of what I now realize was an old dew pond and they survived without undue coddling. In these days I was more interested in the dogs finding hedgehogs in the hollow trunk of an ancient willow which leaned athwart the Dusty Millers but though I don't know the name of the variety I can still, after half a century, recall their purple blue-shaded petals and mealy leaves silver against the ochreous clay.

Like Hazel, Pussy Willow or Palm will almost certainly be a working part of an old cottage garden, especially in wet low-lying land, but as this well-loved English name is another loose term for a whole range of species, I can do no better than quote R. S. R. Fitter's lucid guide.

'Commonest is the Osier, *Salix viminalis* L., which when pure bred is easily told by its remarkably long (to ten inches) narrow leaves but it hybridises readily with other willows. Both Osier and its hybrids are frequently planted for basket-making, for which their long flexible twigs are especially suited. . . . Several kinds of willow, *Salix*, have fairly broad leaves, among them the wide-spread sallows and three northern willows. Commonest and most wide-spread are the two Sallows known collectively as Pussy Willows, from their golden male catkins or 'palm' which appear before the leaves from late February onwards; the female catkins are dull and green. These are the Great Sallow or Goat Willow *S. caprea* L., and the Common Sallow *S. cinerea* L. which is even commoner in damper spots; both may grow into trees. They can best be told apart by feeling the undersides of their leaves; those of the Great Sallow are always downy and soft to the touch, while the Common Sallow's are never soft and usually have at least a few minute rust-coloured hairs. Another way is to peel the older twigs; those of the Common Sallow have ridges under the bark; those of the Great Sallow are smooth.'

Years ago most Willows were pollarded for the utility of their young shoots and often lived to a great age all gnarled and hollow. Now that hurdle and basket weaving are, like most cottage industries, in decline, the Willows go unpollarded and their withies grow into large branches which catch the wind and fall splitting the old trees. The resulting absence of old pollard Willows is doing as much as Dutch Elm disease to alter the traditional aspect of the East Anglian waterways as known to the Norwich School painters.

Hellebores get their name from the Greek, 'hellein', to kill, and 'bora', food, because the species is poisonous which, of course, by the logic of medieval medicine meant that man or beast were to be dosed with it if possessed by a devil. 'A purgation of Hellebore is good for mad and furious men, for melancholy dull and heavie persons and briefly for all those that are troubled with black cholor and molested with melancholy' advised Gerard following Greek tradition.

Our own natural species, *Helleborus viridis* L., Green Hellebore, and *H. foetidus* L., Stinking Hellebore, are both found in the shade of chalk woodlands in the south of England. *H. foetidus* in spite of its poisonous and extremely purgative character was used as a vermifuge, even though it put the patient more at risk than the worms, so we now find it as a garden escape all over Britain. In spite of its rank smell, the fresh green of its large bells coming as they do in the still leafless woods of early spring make it an attractive plant of the cottage garden. Though the Green Hellebore has its yellowish green flowers often edged with purple, the true Purple Hellebore or Lent Lily, *H. orientalis*, with larger flowers, is an introduced species from the Mediterranean. Just to confuse us further Oriental Hellebores have various green varieties. Though known to Parkinson, these reddish-green Hellebores were mainly nineteenth-century developments. The Christmas Rose, *H. niger*, on the other hand, was known to the Greeks. Though we now grow it for the pleasure of its waxen white flowers in the darkest days of winter, it is named for its black roots of more interest to ancient apothecaries who ground them to powder for taking as snuff. Although we have no record of how long ago *H. niger* was introduced from the mountain regions of Southern Europe it is extremely hardy and long-lived, the best I know having certainly been undisturbed for a century, and probably much longer, under a sheltered wall at Kellie Castle in Fife. An annual mulch from the farmyard ensures large blooms to be picked by the dozen off each plant at Christmas. With its liking for lime rubble and a topdressing of leaf-mould and decayed manure, Christmas Roses have always been favourites for a sheltered site at the cottage door.

Scillas, known as Starry Jacinths, were grown in greater variety in Elizabethan gardens than in those of today. Parkinson, a generation later, could list over a dozen species but nowadays even the once popular introduction from Italy, *Scilla bifolia*, tends to be seen mostly in old gardens. Gerard describes it well: 'Small blew flowers consisting of six little leaves {petals} spread abrode like a star. The seed contained in small round bullets, which are so ponderous or heavie, that they lie trailing

Pussy Willow and (reading clockwise), Grape Hyacinths and Wild Hyacinths, Scillas and Lent Lilies.

upon the ground.' Pink and white forms were once grown but have now given place to *S. sibirica*, with bells of bright Prussian blue, only made popular in the last century.

S. peruviana, mistakenly named, because it derives from Southern Iberia rather than Peru, also has this unusual Prussian blue colour but flowers in a large flattened cone. Although this spectacular bulb was grown here three centuries ago, the only time I have ever seen it was wild on roadside banks near Cape St Vincent.

Muscari comosum, Tasselled Hyacinth, is another unexplained casualty of capricious fashion. Gerard grew it as 'The faire haired Jacinth' in both blue and white forms yet though common in the Mediterranean, we have discarded it in our gardens in favour of *M. botryoides*, Grape Hyacinth, now ousted in turn by a brighter blue species, *M. armeniacum*, from Turkey.

Hyacinthus orientalis, wild parent of our popular spring bulb, comes from Greece via Padua where the Orto Botanico or first botanic garden was begun in 1545. It takes its name from classical association with Hyacinthus, the fair youth, loved by Apollo, and killed by Zephyrus the West Wind out of jealousy when he took the discus thrown in a game by the god and hit his fickle boyfriend on the head with it.

In the eighteenth century, the original species, blue with only about six flowers, was developed by the Dutch into our familiar huge, variously coloured, bloom spikes. The practice of forcing the bulbs to flower early in winter was developed by Martin Triewald, Director of Mechanics to the King of Sweden, from the ideas of Nehemiah Grew who, in 1682, saw that flower buds were already formed within the bulb and suggested 'keeping plants warm and thereby enticing the young lurking flowers to come abroad'.

The Bluebells of Scotland as sung about in popular ballads refer to *Campanula rotundifolia*, the Harebell of English downland, widespread on the braes of glen and moor throughout the summer. In Scotland, the lily which spreads a carpet of blue under the old trees of broad-leaved woodland in spring is called Wild Hyacinth, *Endymion non-scriptus* L. Garcke, perpetuating a case of botanical mistaken identity. When, in the Greek legend, Endymion or Hyacinthus, was accidentally slain, a flower is supposed to have sprung up from his spilt blood inscribed 'Ai Ai', Woe is me, and ever since botanists have been examining the base of flower petals of, among other species, Larkspur, Martagons, various Scillas and Gladiolius, to see which was that chosen. *Endymion non-scriptus* L. means that the worthy Swede had examined it for a sign of the wretched boy and could see nothing written on its bottom.

Because Bluebells were so common and had no part in the pharmaco-paeia they were seldom brought into the Tudor garden. However, given favourable conditions of semi-shade in woodland or overgrown shrubbery they will increase and spread very nicely. Some gardeners dislike this trait and instead of relaxing contentedly to watch this partnership often with Honesty, Welsh Poppy or Wood Lilies, they start 'weeding' Bluebells. The bulbs being brittle resent disturbance and usually break off to disclose the slime used in Elizabethan times to starch ruffs.

In contrast to *Endymion non-scriptus* which is distinguished by its nodding flowers, cylindrical with petal tips recurved and yellow anthers, the more upright garden species, *Endymion hispanicus* (Mill.) Chouard, Spanish Bluebell, has a more open-petalled bell, with its tips not recurved, and purple anthers. It has developed pink, white and pale blue forms but hybridizes freely so, in the three centuries since its introduction, has produced in most old gardens a sort of Bluebell spectrum with variously shaped bells and coloured anthers to the despair of botanists and delight of ordinary folk.

Prunus spinosa L., Blackthorn or Sloe, could be said to be a typical member of the shrubby *Rosaceae*, the Wild Plum and Cherry family of cottage gardens and their sheltering hedgerows, without which they could never have existed in an age before wire fencing. Anyone who has ever attempted to push through a thicket of Blackthorn will testify to its effectiveness as a barrier. The tough branches send out twigs, often at right angles to the main growth, with spines, again at right angles, reminding one in their arbitrary appearance of oriental wood engravings. This habit ensures that Blackthorn knits quickly into a barrier capable of withstanding the strongest winds on the exposed cliffs and ridges where it grows and this resilience renders a well constructed hedge of Blackthorn proof against even hungry sheep determined to invade the garden.

Other benefits conferred by Blackthorn are its confetti-like blossom smothering the twigs during the year's first spell of fine weather, producing as it were white clouds against the blue skies of the spring equinox not to mention a crop of Sloes before the first frosts turn the leaves to pale amber after the autumn equinox.

On the dry heaths of East Anglia where Blackthorn abounds, sloe gin was a favourite old country tipple. I remember as a child being set to prick each hard little berry six times with a darning needle before packing them into bottles layered with brown sugar and that a sip of Granny's twelve-year-old sloe gin was memorably different to anything else I have ever tasted. Now, alas! people who will be fobbed off with alien hedges

of useless cupressus will accept the thin pink stuff sold as sloe gin and Blackthorn doesn't find its natural place in the garden, stopping up a wind-fretted corner as it should, to provide shelter for the first spring bulbs.

Our Wild Daffodil, also known as Lent Lily, *Narcissus pseudonarcissus*, is distinguished by paler primrose petals surrounding its long yellow trumpet. It is an indigenous species once plentiful near London and still to be found wild in the Weald and in the New Forest, in Wessex, on the Welsh Borders and in the Lake District. Wordsworth was an accurate observer and right when he saw that his light wild flowers on short wiry stalks 'danced' in a way that our fleshy cultivars never could.

The name Daffodil comes from 'Affodyl', a Tudor corruption of and confusion with 'Asphodel'. The clumsy Latin *'pseudonarcissus'* means Bastard Daffodil because the long yellow trumpet forms were early recognized as 'wrong' relations of the short-cupped white Narcissus, the true Poet's Narcissus of classical Greece, rather than as anything to do with Asphodel, though the name Daffodil stuck, and even the charming Daffadowndyllyes was common Elizabethan usage for any Narcissus with a long yellow trumpet.

The Tazetta or bunch-flowered type of Narcissus now grown commercially in the Scilly Isles goes back even further to the funereal wreaths of ancient Egypt where they have been found still preserved after three thousand years' entombment. Narcissus was the flower worn by the sleeping Persephone when she was taken by Pluto to be his Queen of the Underworld, which accounts for some having turned to gold at his touch though Ovid's story of the boy Narcissus's self love and consequent metamorphosis was dismissed by Pliny who averred that Narcissus meant a narcotic scent derived 'of Narce which betokeneth nummedness or dulness of sense, and not of the young boy Narcissus, as poets do feign and fable.'

Gerard cites Sophocles on Narcissus: 'the garland of the great infernall goddes, bicause they that are diparted and dulled with death, should woorthily be crowned with a dulling flower.'

The Furies wore Narcissus in their hair and from the times of the classical Renaissance to that of Florence Nightingale the scent of Narcissus was considered harmful so that the night nurse of today who removes jonquils from the sickroom is part of an unbroken tradition stretching back beyond ancient Greek legend to Egypt.

The Tenby Daffodil, *N. obvallaris* Salisb, with six pronounced lobes to its trumpet and long known to have grown in the seaside turf of South

Wild Daffodils dance in the shelter of Blackthorn while a sterile double hybrid has fallen to make a meal for Arionater among Periwinkles.

Wales, is believed by most authorities to be a native species, a possible relic of our old Lusitanian flora including such other local survivals as *Romulea columnae*, the Sand Crocus of Dawlish Warren, *Erica vagans*, the Cornish Heath found on the Lizard, and *Arbutus unedo*, the Strawberry Tree of S. W. Ireland. Iberia not Greece is now considered to be the geographical origin of the genus Narcissus so Tenby could well mark the northern extent of its natural range. *N. obvallaris* is six inches shorter than *N. hispanicus* Gouan, the typical uniformly yellow Daffodil of the garden from which such well known varieties as King Alfred were later developed. An androgynary form, with pistils and stamens developed into petals, resulting in sterility and the so-called double Daffodil has been with us almost as long, as it seems to have occurred naturally and increased by root division.

Other varieties have since been developed, such as 'Codlins and Cream' and 'Butter and Eggs', names evocative of the cottage garden, but all tending to be too heavy for their stalks to hold up in spring showers. A night of rain and round you must go picking up the broken cultivars for there is small chance of their being able to dance themselves dry in the breeze.

'A perfect classification of Narcissus does not exist, though many able botanists have attempted its provision. It should contain clear distinctions between species, provide tested pedigrees of intermediate forms suspected of hybrid origin and be sufficiently simple.' E. A. Bowles, in his monograph on the genus, goes on to demonstrate the well nigh impossibility of tracking down reputed wild species to their native stations let alone sorting the various natural and garden hybrids. Where experts give up in despair there is perhaps a crumb of comfort for us lesser mortals as we have to go on trying however vainly. Even if we can do no more than cite the odd variety about which there seems to be some agreement, it will perhaps suffice in a non-specialist book of this sort to indicate the sort of origins from which our cottage garden Narcissus are believed to have sprung.

Narcissus x biflorus Curtis owes its name to William Curtis who established it scientifically in the *Botanical Magazine* in 1792. However, according to Gerard, it was a common garden plant in the sixteenth century under the name of Common White Daffodil or Primrose Peerless. A non-fruiting hybrid, with only one or two flowers on each stalk, Parkinson describes it as 'Whitish cream colour somewhat . . . of a pale primrose . . . with a small round flat crowne rather than a cup, being of a sweete but stuffing scent.'

Though Clusius referred to it in his *Historia* as spontaneous to

Honesty, purple and white, yields nectar to an awakened Tortoiseshell. Modern Narcissus hybrids derive from the Tenby Daffodil (below).

England, Parkinson could never discover its place of growth and Salisbury, after dissecting over a thousand specimens without finding a seed, called it 'one of Nature's mules'.

N. poeticus was a name first used by L'Obel in his *Stirpium Adversaria Nova* of 1570. Linnaeus made the shape of the corona the distinguishing character in the four one-flowered species. *N. poeticus* must be rotate or wheel-shaped as opposed to the campanulate, bell-shaped, *pseudo-narcissus*; the turbinate, cone-shaped, *bulbocodium* and the short six-cleft cup of *serotinus*. Even in the wild, however, *N. poeticus*, Pheasant's Eye, varies much and for our purposes wild forms can be distinguished as early flowering var. *recurvus*, or as the later, larger-flowered, garden form, which has never been traced to a wild origin.

Lesser Periwinkle, *Vinca minor* L., is a doubtful native of Britain. Its Latin name was formerly *Vinca pervinca* from its ancient name of 'Pervink', of which Periwinkle is a modern corruption, nothing to do with shell-fish. Alice Coats points out that the medieval usage of 'the pervink of perfection' also became corrupted in another way into 'the pink of perfection' and was nothing to do with Dianthus. I had always supposed that to be 'in the pink' signified the colour of one's coat and that one was well enough to be out hunting which shows how far the changing use and misuse of words can take us, running about and taking root like the stolons of *Vinca minor* itself.

The Romans used these runners for plaiting into ceremonial garlands and by the Middle Ages this use had degenerated into the recognized head-dress according to an old rhyme for those on their way to execution,

Crowned with Laurer, hye on his head set

Other with pervink, made for the gibet

In Italy, perhaps because of its evergreen leaves, it was reserved for dead children as 'Fiore di morte'. In France it was 'Violette des Sorciers', recommended pounded with earthworms as a love-potion by Albertus Magnus. Apuleius, according to his *Herbarium* of 1480 used it against demon possession, homeopaths use a decoction of it for painful periods in women and Lys de Bray puts fresh leaves up her nose as a cure for bleeding, altogether a more formidable reputation than is suggested by our modern nurserymen's dismissal of it as mere 'ground cover for shady situations'. In the eighteenth century quite a few varieties of Periwinkle were in cultivation; double or single in shades of white, red, purple and blue, some with variegated leaves. Now these seem mostly to have been lost to popular taste, except for *Vinca major* and white or wine coloured forms of *Vinca minor*.

Pheasant's Eye Narcissus, Wild Tulips, Narrow-leaved Lungwort and Soldiers and Sailors.

Alice Coats points out that the old saying 'in cruce salus', health in the Cross, as well as its religious significance for the next world relates equally in this one to the family of *Cruciferae* because all of these simple, four-petalled, cross-shaped flowers, including all cresses, are good to eat.

Lunaria biennis L., Honesty, was certainly well-established here under a dozen different names by the time Gerard wrote his *Herball*: 'The stalks are laden with many flowers . . . of a purple colour, which being fallen, the seede cometh foorth conteined in a flat thinne cod, with a sharp point or prickle at one end, in fashion of the moon, and somewhat blackish. This cod is composed of three filmes or skins, whereof the two outermost are of an overworne ashe colour, and the innermost . . . is thin and cleere shining, like a shred of white satten newly cut from the peece.' Modern scientific botany can hardly better his description. Though we now grow Honesty for its flower rather than to eat, butterflies are not so considerate. As well as feeding from the flowers, several species of Pieridae, Whites, lay their eggs on Honesty so if you like butterflies but wish to grow cabbages free of caterpillars, Honesty should always be encouraged. For centuries countryfolk have enjoyed the dried seed heads as winter decoration and it is so easy to peel the outer cases and empty the seeds on to a newspaper then shake it into a garden corner. Seedlings soon appear which, if it is wished, may be transplanted before they enlarge into rosettes to flower the following summer. These foliage rosettes in themselves are attractive, and for those who don't want a splash of magenta in their spring garden there is a white form.

Tulips on parade are the epitome of formal bedding so it may come as a surprise to find them included among Britain's wildflowers. *Tulipa sylvestris*, Wild Tulip, is always yellow, and sometimes greenish on the outside, with both leaves and flowers slimmer than cultivated Tulips. Though it is naturalized in many places, usually at the fringe of ornamental woodlands, it is decreasing and was probably never part of the indigenous flora on this side of the English Channel. Whether it was growing here before the red Tulips of the East we have no record, only that in 1554 Ogier Ghiselin de Busbecq, the Imperial Ambassador from the Emperor Ferdinand to Suleiman the Magnificent, saw near Constantinople 'an abundance of flowers everywhere – Narcissus, hyacinths and those which the Turks call *Tulipam* – much to our astonishment, because it was almost midwinter, a season unfriendly to flowers. Greece abounds in narcissus and hyacinths remarkable for their fragrance, which is so strong as to hurt those not used to it; the tulipam, however, have little or no smell, but are admired for the beauty and variety of their colours. The

Old Bizarre Tulips, Wild Bluebells and taller Spanish varieties.

RB 5/82.

Turks pay great attention to the cultivation of flowers, and do not hesitate, although they are far from extravagant, to pay several aspers for one that is beautiful. I received several presents of these flowers, which cost me not a little.'

Busbecq brought back Tulip bulbs to Vienna, although he had already got the name wrong: 'Lalé' is the Turkish or Persian name for the flower. In 1561, the Fugger family were growing it in Augsburg; in 1562 a shipment of Tulips from Constantinople reached Antwerp, thence it got to Holland and to England by 1578, because, in 1582, Richard Hackluyt, nephew of 'the Voyager', was able to confirm that 'within these foure yeares there have been brought into England from Vienna in Austria, divers kinds of flowers called Tulipas, and those and others procured thither a little before from Constantinople by an excellent man called M. Carolus Clusius'.

Clusius became Professor of Botany at Leyden and, alas, when he wouldn't sell his Tulips, people came and stole them instead, human nature being less capable of improvement than flowers it seems.

John Tradescant, the elder, grew fifty varieties and Parkinson in 1629 mentions a hundred and forty. Tulipomania of the sort described in Dumas' romantic novel *The Black Tulip* really did take place in Holland in the seventeenth century, with growers paying thousands of florins for the striped Bizarres and varieties such as the famous red and white flowered *Semper Augustus*. Like all fashionable bubbles, it burst and the craze for Tulips returned to something more level-headed. In 1690 the advent of Parrot Tulips was far less remarkable even though the flower painters were quick to appreciate their possibilities. They were even scorned by Van Oosten in *The Dutch Gardener* of 1703 as 'monsters frightful to look upon'. Nowadays, striped or flamed descendants of these once celebrated Dutch varieties are usually to be found, albeit regressed into smaller flowers, almost unregarded, by countless cottage garden paths.

Flowering at the same time and in similar situations we have Lungwort. Soldiers and Sailors is the familiar name of *Pulmonaria officinalis* L. because of its engaging habit of producing pink flowers which then turn blue in a day or two as more buds open so that one sees both colours together on the same shoot. It has, however, been a garden plant long before the days of military and naval uniforms. 'Pulmonaria' means Lungwort and its ancient use for pulmonary complaints was probably suggested by its white blotched leaves, according to the Doctrine of Signatures, and led to its medieval introduction from Central Europe. Today it is a common garden escape whereas our own native species, *P. longifolia* (Bast) Boreau, a taller plant which only sometimes has blotches on its longer, narrower leaves, is now to be found rarely in woods on clay soil in Wessex.

Fritillaria imperialis L., the Crown Imperial, was first grown in the west in the Imperial Gardens at Vienna where it was introduced by Clusius in 1576 as the Persian Lily. However, it is now believed to have originated still further east, in the Himalayas. Though such lush and attractive green growth appearing so early in spring to survive in a hungry world must needs be poisonous and smells of foxes – hence the name Stink Lily – the bulbs 'as great as a pretty bigge childe's head' are edible when cooked and are used in Persian hotpots in the same way as we use the similarly poisonous fungus Gyromitra, rendered harmless by cooking, to season pasta.

By the end of the eighteenth century there were a dozen varieties in cultivation, most of which are now lost to us. This a pity because Crown Imperials will grow easily in the leafmould and dappled shade of light woodland and in the old-fashioned, granny's gardens. And the nectaries of Crown Imperials are as fascinating to children today as they were for Gerard: 'In the bottom of each of these bels there is placed six drops of most cleare shining sweet water, in taste like sugar, resembling in shew faire orient pearles; the whiche drops if you take away, there do immediatly appeare the like, notwithstanding if they may be suffered to stand still in the floure according to his owne nature, they will never fall away, no not if you strike the plant untill it be broken.' I hope he wasn't a rough child: a carefully licked finger tip is enough and doesn't harm the flower.

Cardamine pratensis L., Lady's Smocks to those who went by the Church calendar, or Cuckoo Flower from the season of its flowering, is one of our edible cresses and widespread wherever the grass is less than dry. In the cottage garden it is usually to be found on the drying green simply because dripping sheets and woollens are sufficient to turn growing conditions in the grass underfoot in its favour. The well-drained, weed-free lawns of today's housing estates make Lady's Smocks less common than they should be, which is a pity because in addition to producing a charming flower they are one of the chief food plants of Orange Tip butterflies.

Several of our native rock cresses are extremely rare. Bristol Rock Cress, *Arabis stricta*, lives only on the limestone cliffs of the Cheddar Gorge. Alpine Rock Cress, *Arabis alpina* L., though common in Central Europe

is only found under one rock in the Coolins of Skye. Our cottage garden favourite is *Arabis caucasica*, Willd. from Southern Europe with larger white flowers, which has now settled here so well as to have escaped to join our native flora on limestone rocks in Derbyshire.

Sweet Alison, *Lobularia maritima* (L.) Desv., formerly *Alyssum maritimum* (L.) Lam., from Southern Europe, the familiar edging plant of cottage-garden flowerbeds, has established itself as an annual weed of sandy places and waste ground and has also been accepted into our flora. *Alyssum saxatile*, Gold Dust, which flowers in dense flat-topped clusters over long, grey, slightly hairy, leaves was introduced from Crete in 1710. 'A-lysson' is Greek for calming rage and these edible species were supposed to cure hydrophobia. Madwort was an old name arising from this attribute, though even Tudor apothecaries are unusually reticent about the assertion, so perhaps it was always safer to leave them to the bees who delight in such cushions of honey-scented blossom in the days of early spring when the hive's winter stock is depleted and it is at its most vulnerable.

Although the naming of many plants was aimed to honour otherwise forgotten botanists, popular English pronunciation often makes a doubtful compliment. Andreas Dahl was a Swede, so to a purist the first syllable of Dahlia should be pronounced as in 'darling' rather than 'day' but that everyday use now makes the true pronunciation seem affected. Leonard Fuchs is an ordinary German name but one can see how Victorians, always on the lookout for any hint of indelicacy, had to render the name as 'Fewsha' before they could bring the Fuchsias of 1753 into popularity.

Monsieur Aubriet – 'Oh, Brie, eh?' – was a French Jesuit missionary in China. He wouldn't have recognized his plant namesake as 'Or-bree-sha' so one can only smile at his memory as one waits for the mauve Rock Cress to brighten our walls in spring. Because the lilac-blue *Aubrieta olympica* was introduced in 1710 from the mountains of Greece, it is a plant which will put up with any amount of cold but can't abide a hint of water standing round its roots. Thus an old limestone wall is almost its natural habitat and in cottage gardens this 'need to keep its feet dry' was often met to perfection by planting it in an old chimney pot filled with rubble. *A. hendersonii* is a more reddish purple, *A. leichtlinii* is rose-coloured and from these species many hybrids have since been developed.

Various Fritillaries have recently been introduced from the Mediterranean and Middle Eastern deserts but *Fritillaria meleagris* L., Snakeshead Fritillary, a native flower of damp meadows, is now found wild in only a handful of strictly preserved pastures in Suffolk and the Thames Valley,

The Crown Imperial Lily is spurned by a Peacock in favour of Rock Cresses, White Alison and Aubretia.

all that remain of several dozen places where it grew until the subsidized land drainage schemes of post-war years. There is no need for it to be so rare. Formerly it often found sanctuary in an undrained corner of a cottage holding along with Lady's Smocks, Marsh Marigolds and Rushes. I have put them all in together to remind us of the variety of different plant habitats often offered within the same cottage garden patch.

The Soft Rush, *Juncus effusus* L., is still common enough, though it has lost its former usefulness as the cottage light. Rushes needed to be picked, dried and peeled carefully to extract the pith. This, floated in tallow, burned steadily in a little cruse lamp to supply the only artificial light available to the poor until the coming of oil lamps.

Caltha palustris L., Marsh Marigold, Kingcup, or Molly Blobs in the west, is another plant of the waterside and damp meadows. When cottages were usually dependent on a nearby spring for their water supply Marsh Marigold formed an integral part of the cottage garden and was also more common than it is today. There is a double form sometimes occurring naturally which some people encourage for use in bog gardens: to an artist's eye this looks a mess of busy petals compared to the simple perfection of the single chalice.

Speedwell, species of *Veronica*, I include as a cottage-garden flower, though even Fitter dismisses it as 'a tiresome lawn and garden weed'. How shameful when, on any fine morning in May, a child can see that where a week before all had been an unremarkable dull green, a cloud of sky blue flowers lies lightly on the grass, as magic as a piece of summer sky. Grown-ups are reminded to fetch weed killer and to cut the lawn but fault lies not in Speedwell so much as in our own lost innocence of vision. It is we who have grown too far away from the beguiling simplicity of tiny blue flowers in grass. We can never, alas, go back. There is for us no bottle labelled 'drink me' to open the door to wonderland again. The way lies on through the poetry of Shelley, Wordsworth and Blake to evaluate what we have lost and by using adult discernment at least to learn more about Speedwells.

Both of what are now our two commonest species were introduced from the East. The blue cloud resting on dew-damp grass in May is likely to be round-leaved Speedwell, *Veronica filiforous* Smith. Persian Speedwell, *V. persica* Poir. (V. Buxbaum, Ten.), has larger bluer flowers on longer pedicels or stalklets and is likely to be found flowering at any time of the year rather than in one springtime burst. *V. beccabunga*, Brooklime, is a larger, deep-blue flowered waterplant, very pretty for those able to grow it. Some of the others are rare and difficult to identify. Three are

From the watermeadows, a Damsel fly on a soft rush and (reading clockwise) Snakeshead Fritillaries, Kingcups, Speedwell and Lady's Smock.

tiny specialities of Breckland heath: to sort them out means searching for rabbit holes where the plants are less dried up, and diving down, bottoms up, lens at the ready, in the company of erudite ladies and gentlemen of the Botanical Society of the British Isles and trying to decide whether a fruit is longer than broad or broader than long, whether leaves often measuring no more than a few millimetres across are 'conspicuously dentate' (toothed), have 'finger-like lobes' or have merely been scratched by a rabbit. On the other hand I am grateful to our own Highland speciality, *Veronica fruticans*, for being so rare and shy to flower on the high and remote cliffs where it grows that it has taught me an improved sketching technique. If you stand on one leg and use one hand to hold on by, you can brace yourself with your knee and wedge your sketch pad under your chin like a violin to draw such an Alpine species in situ. Don't look down, and hurry before you get cramp, but the exercise should prove memorable and perhaps, after half a century, enable you to regain something of the intensity of vision you lost on a Speedwell-spangled cottage lawn at the age of three.

WHITSUNTIDE

Wood Lilies for Shady Corners and the Flowers of Old Orchards

Lily of the Valley, *Convallaria majalis* L., is so called from 'Convallaria', meaning from the valley in Latin. Its old country name of Mugget is corrupted French 'Muguet'. The cool restfulness of our Wood Lilies, predominantly green and white, is today unappreciated by an age which usually aims at a crude 'blaze of colour' in the garden. When purple, red and yellow flowers, albeit with no scent, are readily available and easily grown in any featureless patch, fewer gardeners can or will take time and trouble to establish the right conditions of semi-shade demanded by Wood Lilies. Perhaps in Italy, where there is greater need for refuge from the sun-baked heat, one now finds a truer appreciation of the cool green of the 'giardino inglese'.

It wasn't always so. In Jane Austen's day, arbours faced north, woodland walks and shrubberies were designed to offer protection from the sun and only passed out of vogue as fashion came to reflect urbanization. In other ages, when sunbaked peasants toiling in the fields were the norm, gentlefolk took pride in white skins and elaborate clothes. Nowadays when most people are pale in tidy clothes and sitting in offices, fashion decrees that to make an effect you must look as though you've just stepped casually off a Mediterranean fishing boat. The retiring understatement of Wood Lilies is alien to the spirit of the age. Lilies of the Valley take time to grow, resent disturbance and don't photograph well. What could be more damning? Even as a scent, Lily of the Valley has slipped from the high esteem in which it was formerly held. 'Aqua aurea' was the precious toilet water of the Middle Ages to be preserved in vessels of silver and gold. It's no good trying to commend its delicate fragrance to those for whom glamour means tropic palms or a desert ranch. Though once it was plentiful in the Home Counties and round London and was still to be found near Highgate in 1800, Lily of the Valley now survives wild in only a few scattered woods mainly in the less frequented north-west, and every now and then where helpful new brooms have swept clean an old garden they are to be found on rubbish tips.

The name Polygonatum is Greek: 'poly', many, 'gonu', knee joints – Dioscorides' description of the appearance of the rhizome of *Polygonatum multiflorum* L. The rhizome, when cut across, was supposed by some to suggest Hebrew characters and hence the English name of Solomon's Seal but this is discounted by Gerard who says that the plant was used for sealing bone fractures by the 'vulgar sort of people of Hampshire'. There Solomon's Seal grows naturally in the shady beech hangers of the Downs and was brought into the garden under the name of Scalacely in the later Middle Ages. Gerard recommended that the root, pulverized and drunk in ale, 'soddereth and gleweth together the bones in a very short space

Wood Lilies; shaded by Solomon's Seal are Lilies of the Valley and Star of Bethlehem.

and very strangely', also that 'the roots of Solomon's Seal, stamped while it is fresh and greene, and applied, taketh away in one night or two at the most, any bruse, blacke or blue spots gotten by fals or women's wilfulness, in stumbling upon their hastie husband's fists, or such like'. A bolder man than I might comment on that. Considering the plant's long history, it seems a pity that it should have fallen from favour. In *Flora Historica*, 1824, Phillips quotes the Turkish custom of eating the roots and shoots as a vegetable and in the eighteenth century the foliage was esteemed for its floral symmetry. A scented species, Lesser Solomon's Seal, *P. odoratum* (Mill.) Druce, was also cultivated formerly and is still to be found in a few places in the Lake District and the Severn basin. Its flowers are straight-sided rather than constricted in the middle as in the other two species, *P. multiflorum* (L.) Allioni, the Common, and *P. verticillatum* (L.) Allioni, the Whorled, which has thinner leaves growing in a spiral and is restricted to a few woodland sites in Strathtay.

Although I have seen the May Lily in more than one garden, it is a rarity of Yorkshire limestone which, like the more showy chocolate and yellow Lady's Slipper Orchids, should never have been touched. The latter now verge so nearly on extinction that it is possible that insufficient flowers remain in the last surviving site to ensure cross-pollination and the former species would have gone the same way were its quiet, heart-shaped leaves and insignificant fragrant flowers considered so covetously. Our remaining Wood Lilies comprise three species: *Ornithogalum nutans* L., the Drooping Star of Bethlehem, with its bell flowers hanging in a nodding raceme, was introduced from the Mediterranean in Tudor times and still survives in a few sheltered places in the south, and *O. pyrenaicum* L., Spiked Star of Bethlehem or Bath Asparagus, of Mendip woodland, which produces its tall greenish-yellow bloom spikes after the leaves have wilted. For such a large Lily it is amazingly inconspicuous which, as it is edible, is just as well as little boys still gather it for sale in Somerset markets. *O. umbellatum* L., Star of Bethlehem, held to be native possibly in East Anglia, is the common cottage garden favourite. Grass-like leaves come and go before the spreading umbels of jade-green and white star flowers open to the May sunshine. As this species grows only about nine inches high it makes an ideal edging plant for any part of the garden which can be left alone and not dug over relentlessly. Perhaps because of its East Anglian origins, it likes it drier than most Wood Lilies so will seed happily in a sunny border or adjacent gravel path.

Anemones are part of the large tribe of Buttercups, *Ranunculaceae*. Crown Anemone, *Anemone coronaria* L., has been grown in our cottage gardens for a long time, doing well on light loamy soils in any warm corner. Nowadays more vigorous strains from France (Caen) and Ireland (St Brigid) produce larger flowers in stronger colours than the older 'Dutch' Anemones which had really originated in the Mediterranean. The elliptic, overlapping petals, red, pink, blue or white, vary in number from five to eight and the stem leaves, deeply cut into segments and twisted like heraldic mantling, give Crown Anemones great character.

A. hortensis L., is distinguished by its narrow, lance-shaped stem leaves, often entire and bract-like, as well as by its flowers of stronger rose-purple. Though less seen in gardens than formerly, because of their early blooming toughness and intensity of colour, Anemones have retained their popularity as cut flowers and are still to be bought off a barrow throughout the winter.

The more delicate Fair Maids of France, *Ranunculus aconitifolius flore pleno*, are believed to have come here with Huguenot refugees after the Massacre of St Bartholomew in 1572. In the wild plant originating in Bohemia the masses of little pompom flowers are white, which led to confusion with double daisies and the several species known as Bachelors' Buttons. It is, however, related to Buttercups, under the family heading *Ranunculaceae*, so its tuberous roots like Asparagus need to be protected by a sandy/peaty loam from excess moisture in winter and from drying out in summer. It was a rarity to Gerard writing in 1596 who called it Double White Crowfoot.

To confuse identification, both Buttercups and Celandine also produce double forms which have both been referred to as Bachelors' Buttons. The name Ranunculus is a diminutive of 'Rana', frog, hinting surely at Caltha, Marsh Marigolds, yet another flower candidate for Bachelors' Buttons.

Ranunculus asiaticus, Persian Kingcup or Turban Ranunculus, also called Fair Maids of France, came originally from Turkey in the sixteenth century. Gerard, Clusius and Parkinson all had trouble with it whether from corms drying out on the journey, being stolen or merely proving to be single flowered when bought as doubles. Red and white forms seem to have been original, yellow and pink varieties being soon either developed by Dutch breeders or imported successfully from the Aegean, till by the eighteenth century the number of varieties vied with those of the Carnation itself. As many as eight hundred were named on James Maddock's nursery-garden list published in 1792. Now we can only gain some idea of this lost variety from flower paintings and prints because the flower has fallen so far from its former high fashion as one of the seven Florists' flowers that we are offered no more than a 'mixed selection'.

Difficulty of cultivation is held to account for this, though Ranunculus

are no more difficult to grow than Asparagus or Dahlias and a lot less so than many Orchids. It is hard to understand when few flowers are more evocative of the age of the flower prints of Ehret and Redouté and those elegant flower borders of the cottage ornée.

Pheasant's Eye, *Adonis annua* L., is a rare and decreasing wildflower of arable chalkland now on the danger list of species threatened with extinction. With its little red button eye flowers nestling in feathery foliage it is an attractive annual, an obvious relation of the Anemone family and once plentiful enough to be sold in Covent Garden.

As a boy, I remember it in Hampshire, growing with Cornflowers and Ground Pine, now similar rarities, in an odd corner on the way to the cricket field at my prep school. The year was 1940 and the summer that began with wandering along cuckoo-haunted lanes starred with Stitchwort and Buttercups was to end under the whining skies of the Battle of Britain but at least I could still take wildflowers for granted. Since those days a new revolution in farming practice has swept so many of our wildflowers to the verge of extinction, it seems a pity that the old cottage gardening tradition of making room for homely flowers, despised in grander contexts, cannot keep species such as Pheasant's Eye in being.

Welsh Poppies, Columbines and Bluebells are typical cottage garden flowers which delight in such half shady conditions and will quickly clothe as large an area as they are allowed in cool blue and yellow. I suspect that it is just because they are so common and easy to grow that the 'keep control on the treadmill' school of gardening despises them in favour of grey-leaved aliens sold as ground cover. Welsh Poppy, *Mecanopsis cambrica* (L.) Vig., the only species of its genus to be found growing outside Asia, the home of all our celebrated modern Mecanopsis, was cited by Parkinson in 1640 as growing 'in many places of Wales'. It is also native in Wessex and Ireland, and has now spread in gardens throughout Britain. If only it weren't such a good 'doer' it would have more friends. The great Victorian gardener, William Robinson, recognized it as 'the most charming of plants for the wild garden' so as all my gardening is a bit on the wild side, I can love it unreservedly, pulling up seedlings only where I know they will be in the way.

Properly speaking, Apple blossom and Pear blossom should be outside the scope of cottage flowers but as they are there, like butterflies, bees and birdsong, an integral part of the cottage garden with their pink and white petals floating down in the late spring sunshine, it seems unduly pedantic to omit them, especially as the conditions of partial shade occasioned by fruit trees dotted at random in the cottage garden form as important a contribution to its sheltered microclimate as to its general

Varieties of Fair Maids of France, Wild Pheasant's
Eye and Bachelor's Buttons.

romantic character. Old Pears especially, by their long-lived hardiness, ensure that the best varieties available in the eighteenth century still produce their crop of green bullets unappreciated in the spoilt twentieth century when air freight and modern marketing has made us contemptuous of dried and stewed pears. Old fruit trees are a prime target for garden improvers but I use mine as thirty-feet climbing frames for Himalayan briars, so I have put them in to mark my indebtedness.

Aquilegia vulgaris L., Columbine or Granny's Bonnets, a native species of calcareous woodland and its grassy fringes in quiet corners throughout Europe, has been grown in gardens since the Middle Ages. Though poisonous is was used against measles, smallpox and jaundice. According to Alice Coats, Columbines are mentioned in poetry *c.* 1310, as a plague remedy in 1373, as a garnish for 'Gely' in cooking in 1494 and in heraldry as the crest of the Barony of Grey of Vitten. Guillim's Display of Heraldry 1632 refers to its as 'pleasing to the eie, as well in respect of the seemly (and not vulgar) shape, as in regard to the Azurie colour thereof'. John Clare, the observant countryman poet bred on Northamptonshire limestone, recorded its character:

The Columbine, stone-blue or deep night-brown
Their honey-comb-like flowers hanging down.

It remained for Linnaeus to realize that these curiously horned petals, like ruffles of granny's bonnets, were really the flower's nectaries.

Today, gardeners plant the long-spurred Californian hybrids introduced a century ago in place of the dumpy, dowdier Granny's Bonnets but bumblebee pollinators are not so nice and within a few years these flame-coloured exotics can be relied on to hybridize and revert to the homely shades of the old-fashioned favourites of the cottage garden.

Irises are among the most ancient of garden plants, being depicted on the walls of the botanical chamber in the temple of Karnak in 1500 BC and used as 'orris-root' in toilet waters since such civilized notions of cleanliness began. *Iris florentina* L., White Iris, sometimes called Fleur de Lys, is the best species for this purpose, its knobby rhizomes being lifted and dried, then cut in slices and threaded on strings to air completely. Only after two years is the fragrance fully developed. Even if no longer used for a 'pimpled and saucie face' as the Elizabethans recommended, powdered orris, reminiscent of Sweet Violets, still forms the basis of potpourri and a burning perfume for rooms today.

I. florentina is now held to be merely a sub-species of *I. germanica* L., the Common Blue Iris, the purple-blue 'yreos' of medieval manuscripts, which were found distributed in gardens along the caravan routes from

Granny's Bonnets and Welsh Poppies in the shade of an old Pear tree.

India to Persia and the Mediterranean. They are recorded first in Germany grown by Walafrid Strabo, Abbot of Reichenau, in the ninth century. Their rhizomes are almost indestructible and often thrive on rubbish tips after being discarded by gardeners who can no longer be bothered to keep them free of weeds. It is usually a matter of drainage: an iris which never does well, choked with ground elder in the shade of a tree, will flourish wild on the chalk of a railway embankment. On the dry terrace walls of Tuscan farms, Irises will even compete successfully with nettles and, to my mind, the finest Iris garden of all is that in Florence where many species of Iris drift as if naturally among the gnarled trunks and silver leaves of an ancient olive grove, on the way up to Piazza Michelangelo with its celebrated view over the city. Irises are the floral badge of the Medici family and the garden can be seen on application to the Italian Garden Club.

Alice Coats disagrees with Oleg Polunin and others over calling the white *Iris florentina* the Fleur de Lys and decides in favour of *I. pseudocorus* L., the Wild Yellow Flag. She cites the story that in the sixth century Charlemagne's son Clovis, being trapped by a stronger force of Goths on a bend in the Rhine, realized that Yellow Flags growing far out in the river indicated a ford sufficiently shallow for his army to escape. Thus it became the badge of the Merovingian kings, to be revived as his blazon during the Crusades by Louis VII and becoming known as Fleur de Louis – by corruption Fleur de Lys. No other species of Iris, Daffodil or Lily could possibly fit in with this story so the transferred attribution probably dates from the fifteenth-century French excursions into Italy.

In addition to toilet water to clear the skin of spots and freckles, Iris was used for dropsy by Gerard and as a sedative by Turner 'good for gnawings in the belly, and . . . for them that have taken a thorowe cold'.

Because of its colours it is named, like the pupil of an eye, after the rainbow, down which the Goddess Iris, Juno's messenger, came down from Olympus on heavenly errands.

The origin of *Iberis amara* L., Candytuft or Spanish Tufts, is open to discussion: Iberis means 'of Spain', Candia is the old capital of Crete, so you can take your choice over the probable origin of this Mediterranean flower. Gerard got his seed from Lord Edward Zouche on his return from foreign parts and ate it with enjoyment following Dioscorides' practice of using Candytuft seeds as a condiment in place of mustard.

Notwithstanding this introduction, Candytuft is now recognized as a British native species, 'probably' giving way to 'almost certainly' in the latest map in the Botanical Atlas with a convincingly regular distribution

Irises, white Florentine and German, under Apple blossom.

map on the chalk downs along the Ridgeway. Why not? Candytuft belongs to the cresses and is one of the earliest annuals to grow on chalky soils so it probably survived in the wild with the same good-tempered indestructibility as it survived my first efforts at gardening when I used it as a grave flower for the dead canary. Unfortunately morbid curiosity got the better of gardening zeal and I kept exhuming the corpse to see how it was getting on. In spite of this disturbed start on a mulch of yellow feathers and a union jack shroud, as soon as my return to school brought it a short respite, Candytuft spread its lilac and white flowers as a patch of half mourning over my garden.

I. sempervirens, Evergreen Candytuft, came later from Persia via Southern Europe and was grown in the Oxford Botanic Garden by 1679 and in the Chelsea Physic Garden in 1739. The flowers though individually insignificant grow in flat tufts so closely that they can completely obscure the foliage, looking from a distance like a handkerchief 'washed whiter than white' spread to dry over the low shrubbery plant at the edge of a cottage garden path.

Water Forget-me-not, *Myosotis scorpioides* L., is common on wet ground but our ten differing natural species of Forget-me-not were only recognized as such in the nineteenth century. *Myosotis* means 'mouse ear', because of its leaves, softly hairy to deter marauding ants, and *scorpioides* refers to its old name of Scorpion Grass because of the little curled tail of unopened buds, so by the theory of the Doctrine of Signatures formerly pursued by apothecaries, Forget-me-not was supposed to cure a scorpion's sting.

The hybrids we know suitable for spring bedding, were only developed late in the last century but the association of the little blue flower with loving constancy is old. I quote Alice Coats in full.

'About the year 1390, Henry of Lancaster, Earl of Derby and afterwards Henry IV, took as one of his emblems the flowers of *Soveigne vous de moy* or Forget-me-not then credited with the property that those who wore it would never be forgotten by their lovers. From that date onward it appears from time to time in his household accounts in such items as 300 leaves [? flowers] *"de souveine vous demoy"* in silver-gilt to adorn one of his robes or a collar of S's with flowers of *"Soveigne vous de moy"* hanging and enamelled. This stood for Soveignez or Souveraine, either of which words were sometimes worked into the design and the collar of S's, said to have been devised by Henry himself, was worn by both men and women of all ranks to indicate Lancastrian sympathies. The prize in a famous joust fought in 1465 between the most accomplished Knights of England and France was a collar of flowers of *"souveigne vous de moy"* and a collar of S's presented by Sir John Alleyn to the City of London in 1525 is still worn by the Lord Mayor.

Cabbage Whites feeding on Wallflower with, (reading clockwise), Wild Yellow Wallflower, Gold of Pleasure, Perennial Candytuft and Forget-Me-Not.

Gillyflower is the first term which one has to understand. Scientists demanding exactitude and poetic romantics being deliberately obscure are equally tiresome over it. 'Giroflier' is the French name for the Clove Tree, *Caryophyllus aromaticus*, and cloves were used to flavour the rough wine or ale of the Middle Ages. Hence the Anglicized corruption 'Gillyflower' was the name applied to any suitable clove-scented flower. Wallflowers, Honesty, Stocks and Marigolds could all be used as 'Sops in Wine', another Elizabethan name, as well as Pinks, and the notion of them as 'July flowers' makes a good red herring because the heat of summer was when you'd most be wanting a good cup anyway.

Wallflowers, *Cheiranthus cheirii* L., are native originally to Southern Europe but have long been grown here for use as gillyflowers or for nosegays, *Cheiranthus* being Greek for handflower. In days before hygiene was understood and plagues were endemic, the carrying of a nosegay or pomander had a more serious purpose than mere sensuous pleasure. One would like to think it sometimes worked. Because they are clove-scented and edible, Wallflowers were known as Wall Gillyflowers or Yellow Stock Gillyflowers and were one of the species used for sweetening wine and used as such here since Tudor times.

For the story of another of the flower's old names, Chevisaunce, meaning 'Provision for comfort', I am indebted to Alice Coats. The Earl of March's daughter Elizabeth, betrothed to the heir of King Robert III of Scotland, fell in love with young Scott of Tushielaw, son of a border chieftain. In the guise of a minstrel he sang beneath her prison window in the Castle of Neidpath, which is hardly in Wallflower country, and suggested escape and elopement. She signified her agreement by dropping a sprig of Wallflower but, alas, in escaping, she failed to fasten the rope ladder securely so fell to her death. Her love, heartbroken, fled the country with only the sprig of Wallflower in his cap — his Chevisaunce.

Francis Bacon recommended planting Wallflowers under the windows of ground-floor parlours and Parkinson in 1629 listed seven species including red and yellow doubles which aberrants, because they are infertile, can only be propagated by cuttings. Much admired for several centuries, the fashion for these doubles, such as the old red Bloody Warrior, has come and gone the way of other forms of labour-intensive cultivation and only the yellow double, rediscovered by the Revd. Harpur-Crewe, has survived outside the gardens of specialists.

Dame's Violet, *Hesperis matronalis* L., sometimes known as Queen's Gillyflower, is an alien species long naturalized in Britain but originating in the Mediterranean and south-eastern Europe. Even if no longer needed for use as a gillyflower it has remained a cottage garden favourite because

Orange Tips on Dame's Violet; while the male sips nectar the dowdy female quietly lays her eggs on the undersides of leaves.

of its scent and habit of blooming early in summer to the delight of butterflies. Though of a softer lilac colour, the flower heads are similar to Honesty's, clusters of simpler four-petalled blossoms which may almost as often be white as purple.

In the same way it is monocarpic, or biennial, and an observant gardener soon gets into the habit of keeping a lookout for seedlings growing in the wrong place and transplanting them to flower in an odd corner because, again like Honesty, it is difficult to see how any garden can have too much of such very good plants.

Dianthus is literally 'di-anthos', or flower of Jove, because it was the flower used by the Greeks to make crowning garlands for heroes. Carnation is a corruption of this coronation which in later Christian ages was to be transferred to the worship of the Virgin Mary.

Least of all our British natives is the Childling Pink with pairs of tiny pink flowers on a tiny head wrapped in chafflike bracts. An annual waif of sandy shingle, it is rare and most likely to be met with on deserted railway tracks. For reasons known only to botanical taxonomists and the sort of people who dig holes in the same bit of road every few months, it has been named *Petrohagia nanteuilii*, *Kohlrauschia nanteulii*, *Kohlrauschia prolifera* and *Dianthus prolifera*, all in books printed since 1950.

The Maiden Pink, *Dianthus deltoides*, is found wild in dry grassland from the North Downs to the east-coast links of Scotland. The bright pink, single flowers, a centimetre across, are well toothed, with tiny white spots, and the dark green foliage burnishes purple bronze in a dry summer. Its neat habit of growth makes it a favourite for edging a flagstone path or growing on dry walls today as in gardens of long ago.

The Deptford Pink, *D. armeria*, with small crowded heads of scentless flowers, once common on the downlands round London, has now become so scarce that it is more likely to be looked for and found with certainty on the less intensively developed downs of Normandy.

The Cheddar Pink, *D. gratianopolitanus*, bright pink with strongly scented flowers an inch across, is one of our choicest varieties. Although it is cultivated in many cottage gardens of Wessex, in the wild it is to be found only round about the Cheddar Gorge in Somerset. There, however, it is plentiful, spreading over the limestone rocks so that in June they resemble a spectacular rock garden — all undisturbed within earshot of one of our most popular tourist traps. It is highest praise for the National Trust that they manage to keep both worlds happily apart; car park here, buses there, ice creams, lollies, chips and hamburgers, this way to the view, that way to the Pixie Den, drinks on your right, toilets on your left

Wild Pinks; (reading clockwise from top left), Childling, Deptford, Cheddar, Plumed, Clove and Maiden.

and if one among the thousands who would trample it into extinction really wants to see the Cheddar Pink, he can work out where it is likely to be flowering and walk quietly round the back and ascend by unsigned paths accessible to the frailest old botanist.

The Wild Pink and Wild Carnation, *Dianthus plumarius* and *D. caryophyllus*, are not considered truly indigenous species but most likely to have been introduced with shipments of Caen stone, the favourite building material for important castles and abbeys with owners rich enough to ship it over from Normandy. They have certainly been settled on several ruins on this side of the Channel for centuries. *D. plumarius* has petals cut almost to the middle giving it a really feathery look and *D. caryophyllus*, the Wild Carnation, differs from both Wild Pinks and Cheddar Pinks in having smooth edges rather than a rough feel as you run a finger down the edge of a leaf. It has the strongest scent of all and is the parent of later garden Carnations. Both species would have made excellent gillyflowers and were obviously cultivated for use as such.

From a horticultural point of view the early nineteenth century may be called the 'Golden Age of the Pink'. McIntosh listed in *The Flower Garden*, of 1839, 192 varieties and said 'it is pre-eminently the cottager's flower as it takes less care and skill to cultivate than the carnation and other florists' flowers'.

In view of the wild species' obvious liking for extremely well-drained sites on natural limestone or ruined walls it is odd that early gardeners recommended annual top dressings of manure, treatment guaranteed to produce 'soft' growth, killing Pinks with kindness. Those who made gardens out of bombed sites in the war knew better, setting their Pink cuttings into sand and old mortar rubble held up by whatever broken bricks and stones were at hand.

Space precludes mention of more than a selection of the main types. The oldest known variety, dating from the fourteenth century, is Fenbow's Nutmeg Clove which has been recorded as growing in the same garden since the seventeenth century. It is a small double Pink of dark crimson with a rich clove scent, not unlike the wild Cheddar Pink.

'Sam Barlow' was, Roy Genders says, 'At one time . . . to be seen in every cottage garden though is now rarely seen. Like 'Mrs Sinkins' and so many of the old double pinks it splits its calyx but blooms in profusion, its white flowers having a maroon blotch at the centre and with a penetrating clove perfume'.

Some pinks have dark central zones or pheasant's eyes, some have random flecks and splashes of colour and some an almost grapelike bloom on their wine-dark petals such as 'Avoca', the Old Irish Black Pink.

Old-fashioned Cottage Pinks; (see chart), 1 Sam Barlow, 2 Pheasant's Eye, 3 Irish Black, 4 Paisley, laced, 5 Clove, 6 Queen of Sheba, 7 Mrs Sinkins.

'Queen of Sheba' is a Pink of the Elizabethan Painted Lady type, its serrated petals laced with purple on a white ground give the neat artificial look reminiscent of the painted flowers and embroidery of the period.

'In the same way as the miners of Yorkshire and Derbyshire took up the culture of the pansy, and the cotton workers of Lancashire devoted their hours of leisure to the show Auricula, so did the Scottish weavers of Paisley toil to produce the intricate oriental patterns of their shawls on the flowers of the Pink. Their object was to impart the characteristic of rounded or smooth edges to the flowers, thereby eliminating the serrated petal edges of *D. plumarius*.' Thus Roy Genders explains the origins of the Scottish or Laced Pinks which were developed during the early days when handloom weaving was still a cottage industry. The strikingly contrasting petals of laced Pinks appealed to the weavers' sense of design and their culture flourished for half a century until the introduction of the Jacquard loom in 1840 demoted the craftsman to a factory machine-minder for whom working shifts and pollution made pottering among cottage garden Pinks impossible.

'Mrs Sinkins' itself was raised by the Master of Slough Workhouse and named after his wife. A neat mat of silver foliage makes it an ideal edging plant above which the white scented blooms stand up straight, their fringed petals growing in a curiously rounded cabbage shape. The first gardening job I remember was being given a chip basket and blunt-ended scissors and sent to dead-head my mother's 'Mrs Sinkins'.

'Just cut off the little ballies,' I was told and as a three-year-old made a good thorough job, trimming the edging like a little hedge so, when I produced my basketful expecting praise, I still remember my mother's expression changing to dismay. Nobody had said anything about the colour of the ballies: as well as the brown dead-heads I had made a clean sweep of all the much rounder green buds.

Carnations deriving from *Dianthus caryophyllus* L., Clove Carnation or Gillyflower, are extremely hardy provided their simple needs are understood. They are able to withstand intense cold but will not tolerate even partial shade or damp humidity. They don't want a lot of rich humus and manure but must have limestone with dressings of more lime and potash often supplied in the cottage garden by crumbling mortar and old limewashed walls and the odd shovelful of wood ash from the hearth.

Pliny says Clove Carnations were used as sops in wine in Spain during the first century but some authorities suspect an ancient eastern derivation. Latin 'Caryophyllus' comes from Greek 'Karyophillon' and old Arabic 'Quaranful', all meaning a clove. The flower could have come to us from North Africa in the Middle Ages or even earlier via Provence with the

Old Carnations; Red Clove and (clockwise), Striped
Bizarres, Marbled, Flakes and Edged Picotees.

Roger
Banks 6/80

Norman Conquest. The first yellow self Carnations were given to Gerard by a London merchant from Poland and by the time of Charles II there were hundreds of varieties. In the eighteenth century the Carnation became an acknowledged Florist's flower, its diversity of bloom falling under the main headings of Bizarres or striped, Picotees with petals edged with colour, and Flakes with a curious marbled effect on their petals. Only when the industrial revolution had made glasshouses possible did the Carnation develop tender Perpetual and Malmaison strains outside the scope of this book. For the cottage garden, carnations have always meant the old Clove or Border Carnation.

Dianthus barbatus L., Sweet William, came to us from Southern Europe where it is still commonly found in the Dolomites or in sub-Alpine meadows along with the closely allied *D. carthusianorum* L., Carthusian or Charterhouse Pink, which may point to a medieval monastic introduction. Certainly Sweet Williams were established and common enough to have seed sold at threepence a bushel for the planting of Henry VIII's new garden at Hampton Court in 1533. The name Sweet William is thought to refer to St William of Aquitaine but when the flowers became popular in Scotland in the late eighteenth century, the bloody deeds of William, Duke of Cumberland, the 'Butcher' of Culloden, were still too raw in the memory, so to old Scotch gardeners they were Stinking Billies and have remained so to this day.

Matthiola incana (L.) R. Brown, Stock, is named in honour of Pierandrea Matthioli, 1501–77, Italian botanist and physician to Emperor Maximilian II. This Mediterranean flower is surprisingly also a British native but only just, because the south coast of the Isle of Wight marks the northern extent of its range. There it is, however, firmly established on the chalk cliffs between St Catherine's Point and The Needles, sharing the same limited habitat as the Glanville Fritillary Butterfly. To see this unique speckled amber butterfly on the wing and smell the Sea Stocks in flower round the lighthouse makes an object of pilgrimage as memorable to me as the rose windows of Chartres or the golden-lichened splendours of Santiago de Compostella. I digress not wholly unintentionally because this leads me to ask why I should prefer one flower or artefact to another. One looks for perfection of form, so often the simpler the better, in flowers as in works of art. Rarity certainly adds to one's enjoyment whether it be in the poetry in stone of a great building, or a simple Saxon window, untouched for a thousand years, the short span of Schubert's creative output or a butterfly's few days of summer, a flower facing extinction or just autumn sunshine with the threat of winter ending it.

Sweet Williams or Stinking Billies and some
butterfly pollinators of such flat-topped flowers.

All share the same quintessential beauty, absolute, if subjective. Perfection is a point and everyone has his own criterion. It is of no use to tell me that the leaves of stocks look like old rubbed suede, that the stalk or 'stock' itself is often old and gnarled or the flowers ill-formed, I am as hooked on their scent as their moth pollinators and can hang over a bed of stocks by the hour watching them feed.

Such natural interrelationships are surely part of the basis of the cottage garden. All other Stocks descend from *M. incana* except for the Ten-week-Stock which comes from the little wild annual stocks which bloom in the dried up riverbeds of Southern Europe.

The Tudors used them as gillyflowers. 'Dubble Whit Stok Gilliflowers' were purchased for Lord Salisbury in 1611 by John Tradescant the Elder and by the eighteenth century many shades of pink, lilac, and purple variegated with white, single and double, were grown. The celebrated purple and white striped strain of Brompton Stocks owe their name to the nurseries of London & Wise situated west of London in the then little village of that name about a mile beyond where Harrods now stands and whence they were supplied for the Duchess of Marlborough's gardens at Blenheim. Later in the eighteenth century Stocks became another 'Florist's flower', their culture taken up with particular enthusiasm by German weavers.

Stock means in Middle English 'The trunk or stem of a tree as distinct from the root and branches' and to me the gnarled stock is as necessary a part of the plant's appeal as it is in bonsai.

Stock Gillyflowers and Hummingbird Hawk
pollinator of night-scented Sea Stock.

Mock Orange, *Philadelphus coronarius*, was introduced to Europe along with Lilac by Ogier Ghiselin de Busbecq on his return from Turkey to Vienna in 1562. The two shrubs long remained inseparable companions and this led to a confusion in nomenclature that has lasted to this day. At first they were classified together, and shared the name of Syringa, derived from 'Syrinx', a pan-pipe, because the wood of both shrubs, hollow and pithy like that of the Elder, was used by the Turks to make pipes; and we find them together in Gerard's *Herball*, under the names of Blue Pipe and White Pipe Tree. Herein is demonstrated the extraordinary strength of oral tradition; for though the name of Philadelphus was given to the White Pipe by Bauhin as early as 1623, and confirmed for the genus by Linnaeus in 1735, the shrub is still familiarly called Syringa. The exact date of the plant's introduction to Britain is unknown, but Gerard had it in his garden in 'very great plentie' by 1597. Alice Coats' succinctness cannot be bettered. As 'Mock Orange' because of its scent, this wild species from Turkey continued, till our own century, mercifully free of 'improvements' to the flower at the expense of the scent.

'Philadelphus', as in Philadelphia, is Greek for 'brotherly love' but no one knows now why Bauhin chose it as a name for the genus which includes about forty species, part of the order of Saxifrages.

Lilac on the other hand belongs to *Oleaceae* and has a poor relation in the Privet, and a rich one in the oil-bearing Olive. About thirty species are known. Introduced by the returning Ambassador in 1562, Lilac, *Syringa vulgaris*, was so admired as to be illustrated recognizably in Matthiolus' *Commentarii* of 1565. By 1597, Gerard had several Lilacs, the leaves 'crumpled or turned up like the brimmes of a hat'. By 1659, Sir Thomas Hanmer grew both white and red Lilacs as well as the common blue.

The Persian Lilac, *S. persica*, probably the 'blue jasmine' known to Arabs since AD 800, had reached Europe through the Venetian Ambassador to Constantinople by 1614, but it was only when in 1915 it was found growing wild by Frank N. Mayer in Kansu that 'Persian' Lilac was recognized to have been Chinese in origin and another of the plants, along with rhubarb, peaches and apricots, to have taken the silk road to the West.

Paeonia mascula (L.) Mill, was long held to be a native species since it has grown from time immemorial on Steep Holm in the Bristol Channel and at Winchcombe nearby, but this is now discounted evidence as both sites were early monastic settlements. Since Dioscorides it was known as the Male Paeony and the red *P. officinalis* from Crete was known as the

Old single Paeony of the apothecaries.

RB 5/82

Female Paeony. White and double forms of the latter were introduced from Antwerp into Tudor gardens. Because the seeds of some Paeonies are phosphorescent and shine in the dark, the plant became a focus for all sorts of superstitions and healing folklore and was used as a treatment for epilepsy, nightmares and witchcraft until Stuart times. Paeon was a Greek god of healing, sometimes identified with Apollo, hence hymns in his praise were also 'paeans'.

Pre-eminent among Lilies, Madonna Lily, *Lilium candidum* L. is sometimes called the Cottage Lily because it is the most typical flower of cottage gardens. It likes a warm corner of rich well-drained leafmould. In the cottage garden this often means growing them by the back door or privy as Lilies are among the rich foliaged species that tended to profit from having the night slops poured over their roots, the rabbit guts slung at them or, as in one case I knew, the floor sweepings of the village fishmonger sloshed down the back yard with a bucket of water. The peaches, roses and Madonna Lilies thriving in the rich tilth of this sawdust and guts regime were among the finest I have ever seen.

Madonna Lily's history is most ancient, being represented on Cretan vases of the Minoan Period 1600 BC. Alice Coats says that 'Its native country is not certainly known, but is thought to be in the Balkans; a theory which is supported by the discovery near Salonika of a hardier disease-resisting variety which unlike most Madonna Lilies produces abundance of fertile seed. If this theory is correct, it is possible that this flower is a survivor from before the Quaternary Ice Age, which destroyed the plant life of most of the rest of Europe'.

Thus Virgil's mention of it is comparatively recent and its distribution throughout the Roman Empire may be due to the use by the legions of its bulb scales as a corn cure. Madonna Lily's association with the Virgin begins with Bede in the seventh century and continued throughout the Middle Ages, although the name Madonna itself was a nineteenth-century term, when the introduction of other white lilies made distinction necessary.

Thus veneration of the lily as a sacred flower went side by side with medicinal use. As well as curing corns, 'these roots are good to break a boil', said William Lawson. They would 'cure burnings and scaldings without scar', said Culpeper, or 'gleweth together sinews that are cut in sunder', according to Gerard. Only in the last century did physicians decide that a bread poultice would do as well. We can ill afford to laugh at medieval practice when in our age of vaunted scientific understanding, flower arrangers snip the stamens out of Madonna Lilies, castrating them,

Persian Lilac and Mock Orange blossom provide
shrubby cover for nesting Blackbird.

when they are supposed to be celebrating a wedding!

Lilium martagon is such an ancient species that though it is still found growing in the wild from Central Europe to Siberia there is no record of its first introduction to Britain. Its old name of Turks' Cap Lily may point to its being brought back by crusaders, a reminder of the turned-up hat of the Infidel, or it may have always been one of our indigenous wildflowers since glacial times. Turner mentioned a reddish purple lily in 1568. Parkinson gives a description of it in his *Paradisus* and grew the white variety, *L. martagon album* as well.

Gerard knew the Scarlet Turks' Cap Lily, *L. chalcedonicum*, from Greece, the Orange Lily, *L. croceum*, from Bohemia and the little red *L. pomponium* from Southern Europe. It was, however, only as comparatively recently as 1773, that Professor Gowan of Montpelier distinguished *Lilium pyrenaicum* as a separate species from *Lilium pomponium*. Maw, the nineteenth-century Devon botanist, said that *L. pyrenaicum* had been naturalized so long in the hedge banks about South Molton that he considered it probably to be a natural species but in view of its otherwise very limited range in the Pyrenean foothills this is now generally discredited.

The most noticeable feature of *L. pyrenaicum* is that after first being attracted by its yellow flowers, one is repelled from too close an acquaintance by its rank smell. This is strong enough to ensure that the plant was never a close favourite. A century ago Elwes recommended that 'if planted in some half-wild corner of the garden it will thrive well and perhaps be more appreciated than if placed in a prominent position'. Thus unpopularity gained the Lily the conditions it required, the half shade of thin grass or light woodland undisturbed for the seven or eight year period needed for its seedlings to mature into flowering plants. Nowadays, such Stinking Lilies, if not superficially attractive to us, have greater interest to a naturalist as examples of plants pollinated by flies. Perception of their beauty lies in the olfactory organs of Diptera which to understand is surely to forgive.

Annual clearance with scythe and billhook of invading bramble and sycamore saplings is quite sufficient to ensure the continuance of ideal conditions for the spreading of Turks' Cap Lilies. It is better not even to rake up this cut-down herbage but to leave it where it lies or pack it loosely over the Lilies as winter protection for the sprouting of bulbs. My only cautionary word is on tomtits. After flowering the lilies produce a magnificent seedhead of club-shaped fruits standing out from the stalk in a way that makes even the most honest flower arranging friend stoop for her secateurs. To her you may say 'Hands off' but as soon as your back is

Purple and white Turks Cap Lilies and fly-pollinated Pyrenean Lily.

turned and the seedhead begins to turn sere and even more attractive, sap green to soft primrose striped with chocolate, then the tits will be at it, ripping open each fruit to gobble up your ripened seed as if it were cream on the top of a milk bottle. So choose your moment to harvest lily seeds and bring them into safety for the winter. Scatter them again in spring, raking them into the leafmould so that only a more reasonable proportion will have to feed hungry field mice instead.

Madonna Lily has been cultivated since Minoan Crete.

To omit roses from a book on cottage garden flowers would be, I fear, a degree worse than trying to cram them in so I must make the attempt and trust that in doing it I don't offend hundreds of better informed old rose enthusiasts.

'The wild roses', says R. S. W. Fitter, with tactful understatement, 'are a group for specialists' but, to attempt a comprehensible simplification as a beginning (see it in chart form in an appendix) there are five basic English wild briars: first, Dog Rose, *Rosa canina*, with curved prickles, flowers pink or white, hips red, hairless and egg-shaped, commoner in the South; second, Downy Rose, *R. villosa*, with prickles straight and leaves downy, flowers deep pink, hips bristly and rounded, commoner in the North; third, Sweet Briar, *R. rubiginosa*, prickles curved, leaves downy, flowers deep pink, but hips red, hairless and egg-shaped; fourth, Field Rose, *R. arvensis*, prickles curved, green-stemmed, flowers always white and larger than Dog's, hips smaller than Dog's, recognized by its styles protruding above yellow stamens; fifth, Burnet Rose, *R. pimpinellifolia*, densely spiny, leaves small and rounded, flowers creamy white, purple-black round hips, creeping in large low seaside patches in the North from which all the Scotch Briars derive. Any of these wild briars may have its place in a cottage garden together with their hybrids and varieties.

R. rubiginosa was formerly *R. eglanteria*, Eglantine to Chaucer, Shakespeare and the Elizabethans. The name derives from Latin 'aculeatus', prickly, by way of old French. In spite of its thorns Eglantine was sweetly scented and grown as garden hedges, in wildernesses and for nosegays. By 1629, Parkinson had a double form and in 1728 Batty Langley had a pot-grown red one 'Proper . . . to adorn the ladies' chimneys and perfume the air of their Chambers with its pleasant and most delightful odour'. A variety called Janet's Pride was found in a Cheshire hedge by Lord Penzance as recently as a century ago. Hybrid Penzance Briars are still for sale but unless the Sweetbriar is the seed bearer, not the pollen parent, the distinguishing leaf scent of the Eglantine is lost.

Then there are five introduced 'Ancestor' roses; *R. gallica*, *R. moschata*, *R. damascena*, *R. alba* and *R. centifolia*, from which, bearing in mind the ease with which the genus hybridizes, all our garden roses descend.

Both red and white roses were introduced to our gardens from France early in the Middle Ages. Eleanor of Provence brought the white rose as her badge when she married Henry III in 1236. Her younger son, Edmund, Earl of Lancaster, becoming by marriage, Count of Champagne in 1275, spent most of his time in France and adopted the red rose as his emblem. Thus the badges were established in the family long before the

Old Roses; Maiden's Blush, striped Rosa mundi, blue-leaved Damask and Red Burnet.

celebrated brawl in the Temple Gardens began the Wars of the Roses.

This original European red rose, the semi-double *R. gallica officinalis* or Apothecary's Rose, seems to have come to Edmund of Lancaster through his wife Blanche from an earlier Count of Champagne, Thibaut 'le Chansonnier', who returned from his Crusade in 1240. The *Roman de la Rose* written *c.* 1260, talks of 'roses from the lands of the Saracens'. Thus the culture of *R. gallica officinalis* in the Near East is supposed to reach back beyond the Romans to the Medes and Persians of the twelfth century BC.

Another ancient rose of this group, the striped Rosa Mundi, *R. gallica versicolor*, was traditionally associated with Fair Rosamund, the mistress of Henry II, but as this twelfth century liaison preceded Thibaut 'le Chansonnier' by more than fifty years and we have the first written records of it as a 'new rose' in the seventeenth century, the association probably arose from wishful thinking.

The Musk Rose, *R. moschata*, was the only climbing ancestor which, because it was used for attar of roses, seems to have come along the caravan routes from its native Himalayas at a very early date. It seems to have arrived here via Italy in the sixteenth century and, though too tender to do well here, bloomed more than once a year and was able to hand on this valuable characteristic to its descendants, the Autumn Damasks.

The Damask Rose, *R. damascena*, originally accepted as a species, is now recognized as a hybrid from *R. gallica x R. phoenicea*, a Syrian wild rose. Brought from Damascus, supposedly by returning crusaders, it is now thought to have been introduced to England from Italy in Tudor times, according to Hackluyt 'by Doctour Linaker, King Henry the seventh and King Henry the eighth's Physician'. As Linaker died in 1524 this fits in with the treatise on roses of 1551 by the Spanish Dr Monardes in which he says that the Damasks had only been known in the West for about thirty years.

R. alba, another ancestor rose, is now believed to be an early hybrid of *R. damascena* and a white form of our native Dog Rose. The albas, such as the White Rose of York, later called the Jacobite Rose, *R. alba semiplena*, are certainly ancient denizens of cottage gardens. They can be sometimes pale pink, as in varieties such as 'Cuisse de Nymph' (nymph's thigh) or even deeper rose, 'Cuisse de Nymph Emue', (passionate nymph's thigh). Maiden's Blush is the name usually deemed more seemly for British gardens.

R. centifolia, long thought to be the hundred-leaved — (for 'leaves' understand 'petals') — rose of Theophrastus and Pliny, is now considered to be a garden hybrid of complicated derivation no more than four

Rosa alba, Pink Moss Rose and robin's pincushion.

centuries old. This Cabbage Rose, as we were to call it, has been claimed as both Dutch and French but probably originated in the East. It has been traced back to 1580 but by 1800 became the epitome of romantic roses. It is sterile but produces bud-sports, of which the Moss Rose is one. Definitely recorded in a 1720 catalogue of plants in Leyden Physic Garden, it was listed for sale by a Kensington nurseryman in 1724 and by 1800 had become a favourite with dozens of varieties. The 'moss' of a Moss Rose, is in fact a malformation similar in appearance to a robin's pincushion, 'glandular projections all over the flower stalk and sepals . . . are enlarged and at the apex of each sepal a leafy winged and mossy elongation may be present' as Graham Thomas describes it. On the other hand, the 'moss' of a robin's pincushion, the familiar parasite growing on the stalks of wild briars, is caused, Michael Chinery tells us, by a small solitary wasp, one of the *Cynipidae*. '*Diplopis rosae* is the insect responsible for the familiar robin's pincushion galls. Males are extremely rare and the species like several of its relatives, reproduces almost entirely by parthenogenesis. The mechanisms of these bedeguar gall formations are not fully understood but the essential factor seems to be that the presence of the larvae causes the plants to grow in a particular way and provide abundant food for the larvae.' We can only stand silenced at the frontiers of knowledge as we contemplate this minor miracle of the hedgerow but it is amusing to think that tiny ruby-waisted wasps a quarter of an inch long have evolved such an exquisitely specialized way of life and that our advanced women's libbers and one-parent families bending the welfare state are left millennia behind by the goings on in a robin's pincushion.

HIGH SUMMER

Researching Scotland's wildflowers on a project for The National Trust for Scotland taught me that botanically speaking the Border is meaningless and such a nationalist division of the British flora quite arbitrary. If one must divide a continually shading spectrum the break should be geographical where, after the milder south, the Pennines stand for a colder wetter climate. The Trent marks the end of peaches, mulberries and Cox's Orange Pippins and the beginning of curlew country and mountain flowers as you go 'up-dale' in Yorkshire. To draw attention to the special flowers which have always thrived in these grey, stone-walled cottage gardens we must recognize that gardens a thousand feet up on the spine of England have more in common with those of Highland glens than say those of the Home Counties and south coast.

The northern garden, until our modern love of rockeries, had nothing to do with heather. That is the first misconception to dispose of: heather was confined to acid moors above the tree-line. Like the red deer themselves most of Scotland's mountain flowers are really woodland species which with the disappearance of their ancient Caledonian pine forests have been forced to adapt to a treeless terrain. Secondly, we must make the effort to rid the landscape of our imagination of all those fire-coloured Azalea hybrids and acres of Rhododendrons, especially the ubiquitous *R. ponticum* which has become such a curse in the wetter north-west that in Donegal, for instance, it is resisting the best efforts of even the Irish State Forestry to prevent its swamping the ancient indigenous oak and yew forests of Glenveagh. Before the advent of nineteenth-century shrubberies, the northern garden was an orderly oasis, often walled, of plants suited to its cooler wetter summers, some older species often lingering in old gardens of the North after being superseded by newer varieties in richer more fashionable parts of the country, others originally wildflowers of sub-Alpine woodland still to be found wild north of the Peak District and not again till we reach similar habitats in central Europe.

The Globe Flower, *Trollius europaeus* L., is typical. Its moon yellow orbs, instantly recognizable as a sort of giant buttercup, are to be met with in Welsh hills and the Lakes, from the Peak District and Pennines to the West Highlands where it is common in pastureland. Because its almost spherical flowers are perfectly adapted to keeping out the rain, and only open fully in dry sunny weather, Globe Flowers are difficult of access for insects, their potential pollinators. Species of Haltica, tiny flea beetles, nothing to do with true fleas except in their jumping skill, have got around this by learning to bore into the flowers' nectaries from underneath. So to be sure that your Globe Flower has been successful in setting seed,

Specialities of Northern gardens; Jacob's Ladder, Globe Flower and Flea Beetle pollinators, Wood and Blood Cranesbills, yellow Scotch Briar Roses.

look for the blemish of a little hole bored near the base of the petals. If you gently prize open the 'roof' of interleaved petals, you will find the stamens literally hopping with Haltica and that related minor miracles of creation are taking place within the privacy of the golden sphere of *Trollius europaeus.*

Polemonium caeruleum L., Jacob's Ladder, which takes its name from the shape of tiered pairs of leaves, is another native of the Pennines now rare in the wilds which has been grown as a cottage garden flower since Tudor times. It likes limestone and the cool wet summers of its original home in the Peak District so it appreciates growing at the shady base of an old wall where lime mortar leaches out. It has now spread throughout the North usually in its normal lilac-blue form, though a white variety has persisted in probably less alkaline gardens. Both colours were known to Parkinson in the seventeenth century.

Cranesbills are hardy species of perennial Geranium and nothing to do with the half-hardy cultivated pelargoniums from south Europe which are commonly called Geraniums. Several are European natives which have settled here to become part of our accepted flora and some are indigenous. Among the former species are *Geranium phaeum* L., the Dusky Cranesbill, sometimes called Mourning Widow, a name sadly apt because of its bitter chocolate coloured petals and tolerance of extremely shady situations. Look again more closely, however, before dismissing it as not worth while because like all Cranesbills it is a flower of charming details, furry pink ribbed buds, little crimson bracts, well-toothed five-lobed leaves so elegant in character as to be immediately recognizable on the fourteenth-century carved stone capitals of the chapter house in Southwell Minster. *G. versicolor* L., Streaked Cranesbill, is now common in the south-west and *G. pratense* L., Meadow Cranesbill, is the soft blue native of chalk downland. More or less where this last species peters out, *G. sylvaticum* L., Wood Cranesbill, is the more purple blue native of northern hill districts. Also a species predominantly of northern limestone and north-western coasts is *G. sanguineum* L., Bloody Cranesbill, of such a strong crimson colour as to seem a most unlikely wildflower and be welcomed on the most garish modern rock garden. It has a rare variety, *G. sanguineum* var. *lancastriense* (With.) Druce, Lancastrian Cranesbill, which grows on Walney Island, which is so like a common Bindweed flower in form and colour that I once spent an hour in the rain looking for it along the beach when all the time it was as plentiful as Plantains under the wheels of the caravans in the car park and immediately observable to my wife's sharper eye as she sat in the car awaiting my return.

Pansies; Yellow Mountain, Heartsease and Field above, and some old-fashioned cultivars below.

Pansies, with their funny little 'faces' turned to the sun have become accepted as a characteristic cottage garden flower only within the last hundred and fifty years. The earliest pansies with blotches on were grown in 1839 by William Thompson, Lord Gambier's gardener at Iver in Buckinghamshire. They were developed from *Viola lutea* L., the Mountain Pansy, which is a wild flower of limestone hill pastures in north Britain. Plentiful enough in the Lakeland hills and Pennines, the southern uplands of Scotland and on limestone outcrops in the Highlands, it doesn't occur naturally south of the Peak District and the Welsh hills. 'Lutea' is Latin for yellow but in the wild, purple or purple and yellow forms often occur to be confused with *V. tricolor* L., of more general distribution. Sometimes called Heartsease or Love in Idleness, *V. tricolor* has smaller, shorter-spurred flowers and has been a familiar cottage garden flower since Shakespeare's day. The commonest of all, yet most insignificant because its tiny cream coloured flowers have petals often no longer than its sepals, is the short-spurred *V. arvensis* L., Field Pansy, still to be found creeping through the stubble of even the cleanest fields.

V. lutea was known to Parkinson in the seventeenth century as the Great Yellow Pansy but it was only when it was crossed with *V. tricolor* and developed by Scottish gardeners in the nineteenth century that Pansies became a popular cult and capable of endless diversity. James Grieve, after whom the apple was named, crossed *V. lutea* with *V. cornuta*, introduced from Spain in 1776, to produce Violas, freer flowering and standing up on more even stalks above a tidy mat of basal leaves almost evergreen. Because of their mountainous ancestry both Violets and Pansies prefer growing in gardens of the cooler North in shade from the heat of the sun. In the south of England unless they are given a good mulch one tends to find the poor things in late summer flowerbeds wilting under an onslaught of ants.

Our first Lupins were not blue but yellow or white; the Spanish Violet, *Lupinus luteus* L. and the annual *L. albus* L., both grown as a fodder crop, though the fresh seeds are poisonous, or for green manuring and introduced from the Mediterranean if not by the Romans certainly before Turner describes them in 1568. Lupin seeds were used 'to scoure and cleanse the skin from spots' according to Parkinson, and an ointment was described if ungallantly by the Revd William Hanbury in 1770 as used by ladies 'to smooth the face soften the features and make the few charms they possess a little powerful'.

The first of the American Lupins, *L. perenne*, was brought back from Virginia by John Tradescant the Younger in 1637 but, even if perennial,

Old White Lupin, Dusky Cranesbill, Clustered Bellflower and London Pride.

it was another poor colourless thing. It wasn't till 1792 that Vancouver found the Yellow Tree Lupin, *L. aboreus* Sims, on the Californian coast and *L. nootkatensis* Donn ex Sims from Canada managed to colonize the shingle banks of several Scottish Highland rivers. It was only as late as 1826 that Douglas introduced *L. polyphyllus*, blue ancestor of our modern garden Lupins from British Columbia. Even then the species had to wait almost another century for George Russell in his retirement to cross *L. polyphyllus* with *L. arboreus* to produce the brilliant rainbow spectrum we know as Lupins today.

Astrantia major L., Masterwort, is a flower of sub-Alpine woodland and may have been brought here in mistake for Hellebore, as its black roots and their properties were of more interest to apothecaries than the flowers. Gerard calls it Black Masterwort and classifies it with Hellebores. It was certainly here prior to 1597 though we now classify its starry flowers among the *Umbelliferae*.

Sir J. E. Smith appreciated the flower properly in his *Exotic Botany* of 1805 when he says that it is 'not found in every flaunting flower-garden' but as 'a favourite of the more refined admirers of nature'. Thus Astrantia became a 'must' for the shady walks in the grounds of the cottage ornée. It is hard for a sunbathing age always to remember that in 1800 'taking the air' did not mean exposure to the sun which was considered injurious to the complexion and that a sunbrowned skin was positively vulgar. Summer-houses were built facing north as in William Kent's untouched layout at Rousham, newly introduced evergreen shrubs such as Laurel were used to achieve quasi-Italian shade and, where plantations of hardwoods were sufficiently mature, gentlemen could devise elegant grottoes, ladies could encourage the growth of ferns and together they could indulge poetic sensibilities and polite thoughts of a philosophical nature. Floral candidates for such an Arcadia tended to be a difficulty. The currently fashionable Florists' flowers such as Carnations and Fair Maids wouldn't put up with shade and dripping trees. Polyanthus and Wood Lilies flowered before midsummer and the onset of hot weather in July and August, so species such as Astrantia and Campanula which could be relied on to produce elegant flowers in shady conditions gained a new importance.

Species of Saxifrage known collectively as London Pride were also capable of being pressed into the service of this new gardening ideal. Excluding our rare and as yet undiscovered Alpine Saxifrages there was *Saxifraga spathularis* Brot., St Patrick's Cabbage, from Ireland distinguished by its oval leaves, still seen as natural ground cover in the background of

Masterwort or Astrantia, Peach-leaved Bellflower, and Hart's Tongue Fern.

old policies such as Greenfort, Port Salon in Donegal, just woods, rocks and ferns, a few choice shrubs, a seat with a view over Mulroy Bay and St Patrick's Cabbage flowering like a pink mist at your feet.

S. spathularis has also hybridized with *S. umbrosa* L., Pyrenean Saxifrage, with oval obtusely-toothed leaves and now growing wild in the Pennines, to produce true London Pride, *S. spathularis x umbrosa*, now called *S. x urbium* D. A. Webb, and sterile, to be propagated only by 'runners'. It is this London Pride which thrives in drier conditions and which had become by the eighteenth century a popular plant for edging flowerbeds in the Home Counties.

Out of a dozen or so Campanulas now accepted into our flora several species have been introduced and become garden escapes, though surprisingly by no means the most splendid looking of them. *Campanula glomerata* L., Clustered Bellflower, which produces its strong violet-coloured bells massed on a short stalk so that the sight of an established clump of them in flower is arrestingly exotic, is to be found growing wild in Scotland. Although really a plant of chalk downland such an attractive flower tends to be picked out of existence in its original habitats in the Home Counties north and south of the Thames; it enjoys a less troubled existence on a few Scottish sand-dune sites where drainage is good and shell deposits bring up the calcareous level needed to make it flourish. In the cottage garden this was usually done fortuitously by having the flower border edging the path to the door and adjacent to the vegetable patch. Thus lime loving plants or calcicoles such as Pinks, Flags, London Pride and Campanulas tended to get a double dose, both when the vegetable patch was dug over and limed and natural leaching from a flagstone path or crumbling mortar walls.

Other native species of Campanula grown in cottage gardens since Tudor times are *C. latifolia* L., the Greater Bellflower of northern woodlands and *C. trachelium* L., Nettle-leaved Bellflower, of heavy clay soils and sometimes called Bats in the Belfry.

C. persicifolia L., Peach-leaved Bellflower, and *C. medium* L., Canterbury Bell, now sometimes growing wild as escapes, were originally brought in from Europe in Tudor times. Both blue and white forms of *C. persicifolia* were known in gardens here before 1596. According to Parkinson 'The peach-bels as well as the other may safely be used in gargles and lotions for the mouth, throat and other parts as occasion serveth. The roots of many of them, while they are young, are often eaten in sallets by divers beyond the seas,' that is, a note adds, 'in places where the natives are not particular about their salads and side-dishes'. Personally

I suspect that even in the seventeenth century 'divers' natives overseas were more particular over the variety of their 'sallets' than the insular British, Florentine cookery would suggest so.

In the garden the Peach-leaved Bellflower is more tolerant of semi-shaded situations so tends to survive in overgrown corners of an old garden whereas *C. medium* L., Canterbury Bell, needs full sun on chalky soil and has long been a favourite of gardens in the south of England. It is now believed to have originated in the Pyrenees but Gerard, having described it identifiably with flowers having 'much downie haire such as in the eares of a dogge', goes on to call it Coventry Bells following Lyte's *Niew Herball* of 1578 mentioning that 'they grow in woods, mountains and dark vallies . . . especially about Coventry . . . for there is another kinde of Bell-floure growing in Kent about Canterbury.' Neither can have checked these Coventry Bells or their environs and may have got mixed up with Rampion, *C. rapunculus* L., or even *C. patula* L., Spreading Bellflower, both of which had formerly a strong Midland distribution though they are almost extinct today. Parkinson disproves the error and Hanbury adds that, like Rampions, these unknown Coventry Bells were liable to be grown, boiled and eaten as a vegetable.

Foxglove, *Digitalis purpurea* L., our native woodland species familiar by its graceful spire of purple or white bells, is one of our few wildflowers still used in modern medicine, as the source of the heart drug digitalis. Paradoxically this genuine virtue was only discovered by a Birmingham physician in the late eighteenth century and was quite unsuspected by the old herbalists who believed right or wrong that they had found a use for almost everything. They used Foxgloves in their treatment of skin sores and scrofula from which the sub-order *Scrophulariaceae*, two lipped flowers, takes its name. The name Foxglove comes either 'from Anglo-Saxon Foxes-gleow, the gleow being a musical instrument consisting of an arch supporting a ring of bells of graduated sizes', Alice Coats' explanation or, according to Lys de Bray, from 'bad fairies [who] gave the flowers to the foxes to put on their paws so that they could creep silently on to their prey.'

Otherwise, Foxgloves provide another instance of the contracting tastes, especially in woodland flowers, of today's gardeners. Parkinson grew four species and even Gerard grew *D. lutea*, the Alpine yellow, and *D. ferruginea*, smaller buff-coloured, two of the best European wild Foxgloves. It is sad that such graceful flowers familiar to the Elizabethans should be almost lost to us when they are so easy to grow, requiring only an undisturbed shady corner of a cottage garden in which to seed themselves.

Aristolochia clematitis, L., Birthwort, is another reminder of past medical practices sometimes surviving in a quiet corner of an old garden or naturalized in a few places in the South-East. The shape of its one-lipped, greenish yellow flower is immediately recognizable and, I suspect, suggestive of its former application in the midwife's trade. The medieval apothecaries, however, had got hold of the wrong end of the stick, because the 'babies' in its 'womb' aren't going to come out: the little flies have crawled in and, prevented by hairs from escape, are alas! having to stay in to be eaten, because Aristolochia is carnivorous. Even if Birthwort has been discarded now by the most home-spun, do-it-yourself delivery room, one can keep a children's party amused for ten minutes by looking for flowers that have eaten their fly. Neat dissection with a razor blade by father discloses the secrets of the 'womb' and if anyone wishes to bet on it, grown-ups can cheat with precognition because the lip folds over to close the aperture once the victim is safely ingested.

Borage and our various Comfreys are considered to be introduced species but Alkanet, once called *Anchusa sempervirens* by Linnaeus but reclassified as *Pentaglottis* by Tausch, has always been a native of Wessex

Foxgloves, purple and white, and Birthwort.

woodlands. Often known as Blue-eyed Mary it has been introduced throughout Britain. Its bright blue flowers, really white-eyed with star-shaped bee guides on their five petals, settle happily in any homely woodland corner where its hispid, or slightly bristly growth, isn't scraped out for something more fashionable. On the other hand, *Omphalodes verna* Moench, also variously called Blue-eyed Mary, Bright-eyed Mary or Venus' Navelwort, came here from sub-Alpine woodlands in the seventeenth century. It is another member of *Boraginaceae*, an order including Comfrey, Anchusa and Forget-me-not, and has shorter, hairy-leaved shoots which will soon carpet light woodland and make it gay with blue flowers before the leaves unfold in early spring. The name Venus' Navelwort comes from the shape of its fruit and Latin wishful thinking.

Veratrum album L., known as White False Helleborine, and *V. nigrum* L., Black False Helleborine, are a wonderfully sinister pair from sub-Alpine woodland where they have the added toxicological charm of being easily mistaken for the large Yellow Gentian gathered for use in beneficial cordials. Polunin points out that 'The False Helleborine has leaves arranged in threes and they are hairy beneath, and the roots have a strong unpleasant smell while the leaves of Gentian are opposite, hairless beneath and the roots are almost odourless.' Veratrum was used by the prehistoric tribes of Gaul for tipping poisoned arrows and was well established here by 1568. A third green-flowered species, *V. viride*, Indian Poke, was brought from America by Peter Collinson in 1763.

Confusion with Hellebores arose from Parkinson's *Theatrum Botanicum* in which plants of similar properties, cooling, hot-biting, purging plants are arranged together. Thus Christmas Rose roots were Black Hellebores and Veratrum became known as White Hellebore. All needlessly muddling unless, as Veratrum moved from the armoury to the medicine chest, you were to be the object of a purge.

In the oft-repeated classical tale of Solon's siege of the town of Cirrheus, Solon was successful because he first dammed the river to create a shortage of drinking water, then steeped Veratrum in it before letting it flow again. As a secret weapon, the resultant dysentery anticipated germ warfare by two millennia and as a violent remedy for violent symptoms Veratrum continued to be prescribed for maniacs and epileptics until the age of science, when it was relegated to use as an insecticide. Russians had always used it as a midge poison – its dialect name 'Cameritza' means gnat plant – and John Tradescant saw 'Helebros Albus, enough to load a ship' at the Monastery of St Nicholas near Archangel, when he went on the 'Voiage of Ambassad' in 1618. Even if we don't suffer from tundra

In the shade of unfolding leaves of Veratrum;
Creeping Jenny, Alkanet and Leopard's Bane.

midge swarms, Veratrum should be more popularly grown today for the sake of its foliage, as Gerard with charming accuracy described its leaves, 'folded into pleates like a garment pleated to be laide in a chest'.

Doronicum pardalianches L., Leopard's Bane, is a yellow daisy growing in similar woodland fringe sites in spring. *Pardalianches* is Latin for Leopard Throttler but even if that happens to be your garden's pest problem, old authorities differed on its efficacy as a poison because of inaccurate identification and confusion between Wolfsbane, deadly Aconitum, and the Doronicum of Arab cordial makers. John de Vroede found it 'verie pleasant in taste and comfortable'. Matthiolus, his contemporary, gave it to his dog which then died. Lyte, following classical tradition, declared 'if this harbe . . . be layd by the scorpion then he shall lose his force and be astonied, until such time as he shall happen agayne to touch the leaves of White Elebor, by vertue whereof he cometh to him selfe agayne.' Clearly there is room for further research and experiment. Nowadays might one not substitute feral mink for leopards?

We have seven species of *Lysimachia*, Yellow Pimpernel or Loosestrife, three now considered to have been introduced and four natives. Of these one, *L. thyrsiflora* L., is a waterside plant with tiny flowers and another, *L. nemorum* L., a weak semi-prostrate woodland growth with few flowers on slender stalks. For the purposes of the cottage garden there remain two old favourites, *L. vulgaris* L., which grows in tiered spires of yellow star flowers two or three feet high to form large clumps making an attractive feature of many a quiet flower border or shaded orchard, and *L. nummularia* L., Creeping Jenny, with similar flowers paired on runners at home in damper corners.

Our wild species, Monkshood, *Aconitum anglicum* Stapf., is more usually found in Wessex woodlands, and has leaves of a lighter green and more deeply divided than the Continental forms usually surviving in old gardens or their hybrids, such as the popular Bressingham Blue grown today. Their deepest indigo, distinctly helmeted flowers are similar in shape to those of cream-coloured Wolfsbane, *Aconitum vulparia* L., from sub-Alpine woodlands. Both were introduced in the Middle Ages. The whole plant is poisonous in all its varieties so the efficacy of a strong decoction of it for poisoning bait ensured it a place in the gardens of shepherd and gamekeeper.

The first perennial Delphinium, *D. elatum* L., came from Russia in Delphinium, were brought from Italy before 1551 and used as a cure for lice. Licebane and Stavesacre are old names for *D. staphisagria*, a species now no longer grown. 'Coarsely powdered, they were strewed on children's hair and this never fails', we learn from John Hill's British Herbal of 1756. The remedy was still in general use at the time of the American Civil War though I remember questions about it didn't go down too well with a high-powered radio interviewer who telephoned transatlantic one evening asking me to ad lib on the stranger uses of herbs. We'd all had a good dinner and forgot that due to the time change it was going out on 'Woman's Hour' to a sober afternoon audience. My doctor brother in Victoria B.C. nearly choked with laughter over his breakfast coffee and I still don't know about the incidence of lice in America.

The first perennial Delphinium, *D. elatum* L., came from Russia in the seventeenth century and from its hybrids with late nineteenth century introductions from America our present garden varieties were developed.

Artemisia and Cimicifuga may sound like a pair of romantic heroines from the sort of novel satirized by Jane Austen in Northanger Abbey and may be, as it were, invited up to the drawing room by modern flower arrangers but as Wormwood and Bugbane they have a more homely part in the cottage garden. Bugbane, *Cimicifuga*, came from sub-Alpine woodlands as an all too necessary adjunct of the dubious personal hygiene beneath those silks and velvets of Tudor finery. Wormwood, *Artemisia absinthicum* L., is the cousin of Mugwort, *A. vulgaris* L., the still more homely weed which enlivens dusty waysides when on a grey day the wind blows up the white undersides of its coarse drab leaves. They are all insect repellents and were strewn on the floor in the Middle Ages to deter fleas.

A. absinthicum L., Wormwood, though esteemed in elegant gardens today for its much branched silky foliage is still to be found growing wild on limestone banks near the sea. Its flowers, chains of yellowish balls, have a look of dead mimosa but its foliage is aromatic and when distilled forms the basis of absinthe. According to Apuleius, the first absinthe drinker would appear to have been Artemis, the Goddess Diana herself, introduced to the secret by Chiron the Centaur and so enchanted by its aniseed flavour but she gave her own name to the herb. Though long esteemed both as a stimulant and as a digestive, over-indulgence results in mental deterioration as depicted by Manet in his studies of seedy French café life.

In the same useful family, *A. dracunculus* L. got its name as a dragon bane (keeping newts out of a damp cellar perhaps?). We know it as 'Russian' Tarragon, the perennial herb for use in the kitchen, which came from Asia before 1548.

Finally, with similar virtues as an insect repellent, *A. abrotanum*, Lad's Love, Old Man, Maiden's Ruin or in Scotland, Apple Ringie, has the

most feathery foliage with the sweetest scent of all so tended to be kept for keeping moths out of my lady's closet. It was extolled by Dioscorides and is known to have been grown in England since 1440. As well as the all too frequent medieval recommendation as a wound herb one thing evidently led to another and it acquired a Tudor recommendation as an aphrodisiac which explains a lot of its names and, one would like to think, its place of honour at many a cottage door.

Every nice family is liable to have a skeleton in the cupboard and the *Caryophyllaceae* are no exception. It seems extraordinary that a group including Maiden Pink, Carnation, Campions and delicate Catchfly can include *Saponaria officinalis* L., Soapwort, but there it is. Bouncing Bet and Goodbye to Summer are other names descriptive of this irrepressible hoyden of which Gerard complained 'if they have but once taken good and sure rooting in dry ground it is impossible to destroy them', so the pretty pink *Saponaria* tends to survive in odd corners to delight our autumn days. It was grown for its cleansing properties 'to scour the country-woman's treen and pewter vessels' according to Parkinson but is still recommended by the British Museum's restorers for ancient fabrics too delicate to trust to modern detergents.

Marigold, *Calendula officinalis* L., is literally 'of the Kalends' because it is likely to be found flowering on the first day of every month in the frost-free lands whence it originated. Marigolds or just plain Golds have been popularly grown in Britain since the Middle Ages for both medicinal and culinary use. Pounded with Sage, Sorrel and Tansy among other herbs, then covered in water, Golds were held by Culpeper to be an effective guard against the 'Pestilence' or were applied in a compress for sprains. Dried as a poor man's substitute for saffron, according to Gerard, Marigolds had their place on the kitchen shelf for use in winter stews till quite recently. Husbandman's Dyall was another Elizabethan name deriving from the flower's habit of opening and shutting with the sun and thanks to the plant's happy fecundity there are few old cottage gardens without Marigolds to cheer them in autumn.

Homely remedies from the cottage garden (left to right), Soapwort, Bugbane, Wormwood, Monkshood, Larkspur.

Althea rosea L., Hollyhock, is of ancient introduction. Alice Coats points out that 'hoc' is Anglo-Saxon for a mallow and that, as the plant is common in the Mediterranean, 'Holy-hoc' may indicate a medieval pilgrim origin. John Gardiner's *Feate of Gardening*, 1440, mentions 'Holyhocke'. Turner has it as 'our common holyoke' in his herbal of 1551.

The original colour is a soft pink as in *A. rosea* and the closely allied *A. cannabina*, but Parkinson had 'a darke red like a black cloud' in the seventeenth century, as well as a double, and admired the way in which the tall spires go on producing flowers well into late autumn if the weather is fine. By the end of the century Sir Thomas Hanmer grew a pale yellow and recommended the plant as 'fittest for courts and spatious gardens, being so greate and stately'. Lord Burlington cultivated varieties from China where the flowers were also eaten as a delicacy and a Regency attempt was made to grow Hollyhocks commercially according to Alice Coats 'in order to use the fibres of the stem in the manner of hemp or flax'. Indeed, when Indian Hemp, Cannabis, appeared growing at Hyde Park Corner (whether by chance or design was never ascertained) on a soil tip from the underpass construction in the hot summer of 1955, the more innocent thought it was seedling Hollyhocks. In the last century and a half the plant has fallen from favour, indeed the only Hollyhock readily obtainable is the double, a sterile deformity looking about as attractive as a well-used red paper hanky.

I don't want this to be another book on herbs but how can anyone writing about old cottage garden flowers not include Lavender and Rosemary? Lavender gets its name from the Latin, 'Lavandus', to be washed, so the use of it as toilet water probably dates from Roman times at least. 'Llafant' is mentioned in the thirteenth century *Book of the Physicians of Myddvai*. In 1568 Turner recommended lavender flowers 'quilted in a cappe and daylye worne for all diseases of the head that come of a cold cause' and Gerard prescribed it 'for them that use to swoune much'. Lavender lawns, trimmed with a scythe, existed in some grand gardens in the seventeenth century.

In Provence before the two world wars, according to Alice Coats, 'the lavender still [for distilling!] used to visit all the remote little mountain villages in turn, and the peasants used to bring down their loads of lavender from the hills to be distilled in the market place; the perfume predominating for a time over the usual village smells of drains and garlic. One would expect the plant to secrete its aromatic oil more freely in a hot dry climate but actually Lavender, Rosemary and Peppermint grown in this country produce a more abundant and fragrant essential oil than

Old single Hollyhock, Lavender and Rosemary.

RB 10/81

anywhere else in the world. Three hundred acres of it were formerly grown near Mitcham in Surrey and the oil of lavender produced there realized six times the price of that distilled by French growers.'

Of the species grown, *Lavandula vera* (*L. officinalis* Chaix), now *L. angustifolia* Mill., is our Common Lavender and *L. spica*, the tall Spike Lavender which gave its name to jelly flavoured with it as 'aspic'. Dwarf Lavenders, known to us as Munstead varieties after Gertrude Jekyll's late Victorian garden, were grown in Elizabethan days as well as *L. stoechas*, the tender Pennant Lavender named from its purple bracts above the flower spike which features in a fifteenth-century list of herbs as 'stycadose', taking its Latin name from the Stoechades, Iles d'Hyeres, off the Côte d'Azur near Toulon.

Rosemary, *Rosmarinus officinalis* L., and Lavender, the two most favoured herbs of the cottage garden, often grew on either side of the door. Rosemary takes its name from Latin dew of the sea, 'rosmarinus', because it was known to thrive best near the shores of the Mediterranean and has not changed in two millennia of garden cultivation from the wild shrub of the Maquis.

Like Lavender, Rosemary had countless uses in toilet preparations and was a chief ingredient of Queen of Hungary's water for the hair. In Shakespeare, Ophelia's 'Rosemary for remembrance' is symbolic of classical use of Rosemary as a memory aid for students. Gerard quotes Serapio: 'Rosemary is a remedie against the stuffing of the head that cometh through coldness of the braine, if a garland therof be put about the heade.' Queen Philippa, wife of Edward III, received a manuscript from her mother, the Countess of Hainault, in 1370 in which the virtues of Rosemary are set forth, and Sir Thomas More wrote 'lette it runne all over my garden walls, not onlie because my bees love it but because it is sacred to remembrance and therefore to friendship'.

Sprigs of Rosemary were thus placed in the hands of the dead and thrown into the grave. For the living, Rosemary can be used as substitute incense or in the kitchen, where a few sprigs slipped with garlic near to the bone of roast lamb transforms it.

Another species introduced originally from the Mediterranean and equally suitable today both for the flower border and the kitchen garden is *Nigella damascena* L., so named because of its black seeds which were supposed to have been brought from Damascus in 1570. We know it as Love-in-a-Mist today, looking no further than its pale blue flowers half-hidden in a 'mist' of finely divided foliage, while Parkinson called it Fennel Flower.

Devil-in-a-Bush is a name suggested by the great horned seed heads often turning red in a period of dry weather. The seeds, however, are aromatic and, says Alice Coats, 'contain a slight narcotic principle called "damascinine". (Was this the "Love"?) The seed is still used as a flavouring in France where it is known as "Quatre-épice" or "Tout-épice" and is sometimes ground to adulterate pepper'. Marcel Boulestin suggests that today the seeds lightly crushed in a mortar may take their place with ground cinnamon, coffee and chocolate as an accompaniment of pommel cheese and cream.

Gladiolus illyricus Koch is high on the list of our protected wildflowers. It flowers right in the middle of the tourist season on a handful of secret sites in the New Forest. 'How,' I asked when commissioned to paint it for the Hampshire Naturalists' Trust who safeguard it. 'How does a bright pink gladiolus not announce its whereabouts to every predatory flower picker?'

'Well, you see, it's hidden by the bracken.' I visualized huge spreading fronds of bracken and Gladiolus stiffly upright and thought desperately of cramming it into my five-inch picture postcard and said that I must see it growing. 'You promise not to tell where you've been?' Almost offering to be blindfolded I was led by the doyenne of county botanists down a sandy dell where after a bit of searching – 'The trouble is the wild ponies eat them too' – I was shown wild Gladiolus and now understand its growth habit. The secret lies in its timing. The leaf shoots come up looking nondescript as rushes before the bracken unfurls, then flower just after the bracken leaves have hidden them from view and set seed before they can damp off under the heavy green canopy of late summer. If left alone it can survive.

The name Gladiolus comes from Latin 'gladius', a sword, and its old Lyte's *Niewe Herball* of 1578: 'Coarn flagge or Coarn-gladdyn I have it in my garden.' Type is unspecified but it was used for poultices to draw out splinters or 'drunk in Goats' milk to take away the paine of Colique' according to Gerard, who by then had got hold of two European species, *G. communis* which spreads and *G. segetum* with anthers longer than the filaments in contradistinction to our own *G. illyricus*. This is the species most commonly to be found in Mediterranean vineyards in spring and well established in many gardens of Wessex, such as the garden of Rex Whistler's old house in Salisbury Close where large clumps of it grow in dry turf against the wall absolutely trouble free and surviving a large family of children's games, which is more than one could say of the gross cultivars developed from South African species introduced from the Cape in the last century and now known as Gladioli.

Mallows, apart from a few tiny-flowered nonentities are, surprisingly for all their exotic looks, British native species. Marsh Mallow, *Althaea officinalis* L., and Tree Mallow, *Lavatera arborea* L., it is true, are seldom to be found growing far from the seaside, and always seem to be yearning for Mediterranean sun, but Common Mallow, *Malva sylvestris* L., with folded lobes to its large leaves and Musk Mallow, *Malva moschata* L., both have their place in the cottage garden because, like the celebrated Marsh Mallow to be found wild on East Anglian salt marshes, their roots were candied along with the base of the stalk in spring before it gets too tough. Roger Phillips says that in addition to cough sweets and the oriental sweetmeat known as Turkish Delight it was Roman practice to eat them as a vegetable and Mallows 'have long been used as a soothing agent in treating the inflammation of the skin, eyes and respiratory, gastric and urinary systems. Chewing fresh flowers relieves toothache and crushed in olive oil they relieve bee and wasp stings.'

Musk Mallow has finely cut leaves which are an ornament to the most refined flower border. Both species produce white forms so they can appeal even to those who find that too many late summer flowers come in shades of sweetie-pink.

Growing similarly together with a liking for well-drained chalky banks in the cottage garden are Sages, species of *Salvia*. Common Sage from the Mediterranean, *Salvia officinalis* L., was well established by Gerard's time and used more for throat gargles and in cheese and wine-making than for forcemeat. Our own native Meadow Sage, *S. pratensis* L., the handsome purple-flowered Clary still to be found wild on Midland limestone, was used as an eye wash, confirmed by its name 'Clear-Eye'. *S. glutinosa* L., Jupiter's Distaff, is another species, very sticky and strong smelling, from sub-Alpine woodland which has been here since Tudor times and is sometimes still to be found growing wild in a shady corner of an old garden. Jerusalem Sage, *Phlomis fruticosa* L., is the most handsome with tiers of hooded yellow flowers up its white-felted stalk. The name Jerusalem Sage may point to a crusading introduction and certainly one of the Palestinian Sages inspired the design of the Jewish seven-branched candlestick.

In the same large family of *Labiatae* are the lowly *Stachys lanata*, familiar to most children brought up in country gardens as Lamb's Lugs, and *S. germanica* L., Downy Woundwort, which hints at a dark past before modern stitches and bandages were available to bind up even minor injuries. In the nineteenth century tropical varieties of Salvia were introduced from Mexico. 'Exotic sages have no moderation of their hues,'

Mallows, cut-leaved and common, Wild Gladiolus and Lambs' Lugs.

declared Ruskin, 'There's no colour that gives me such an idea of violence — a sort of rough angry scream' — and most of us who don't don dark glasses to look at those scarlet horrors in municipal flowerbeds would agree.

Scabious gets its name from fifteenth-century treatment of scabies and similar skin diseases. A sixteenth-century herbal gives account for the name of our native Devil's Bit Scabious, *Succisa pratensis* Moench: 'because the rote is blacke and semeth that it is jagged with bytynge, and some say that the devyll had envy at the vertue thereof and bete the rote so far to have destroyed it'.

The first Scabious to be introduced to our gardens was not the popular blue *Scabiosa caucasia*, which today is always put in bunches with Gypsophila and Sweet Peas by sellers of cut flowers, but *S. atropurpurea* L., sometimes called Pincushion Flower. In 1591 Clusius received it from Italy as Indian Scabious, though, in fact, it is a wild species from the Mediterranean. By 1620 William Coys grew it at North Ockenden 'of a delicate redd colour like to redd velvett', and the very dark form known as Blackamoor's Beauty or Mournful Widow enjoyed great popularity. Whether because Scabious is happier in a warmer climate or because an extravagant panoply for mourning widows is more fashionable in Catholic lands, *S. atropurpurea* enjoys more favour today in France as 'Fleur de Veuve', in Italy as 'Fior della vedova' or as 'Saudade' in Portugal where its musky scent is suitably appropriate.

Ox-Eye Daisy, *Chrysanthemum leucanthemum* L., is our native meadow daisy, sometimes called Marguerite after Margaret of Anjou, whose emblem it was, embroidered in a three-flowered device on her robes and those of the ladies of her suite, when she came to England as a fifteen-year-old bride for Henry VI. Though ousted from garden cultivation since the nineteenth century by the larger more robust *C. maximum*, the big Moon Daisy of the Pyrenees and parent of modern Shasta Daisies, Marguerites will sooner or later make their appearance on any grassy bank that is scythed for hay rather than allowed to go to waste under a mowing machine.

Our other old Chrysanthemums were yellow annuals, the native Corn Marigold, *C. segetum* L., still holding its own in the wild as a weed of arable fields and parent of several garden varieties and the Crown Daisy, *C. coronarium* L., introduced from the Mediterranean prior to 1629 and known to Parkinson as 'Corne Marigold of Candy' (Crete).

This latter has similar golden flower heads to Corn Marigold's but in

Ox-Eye Daisies, Red Hot Poker, Jerusalem Sage and old Dark Scabious.

place of its coarsely toothed fleshy leaves it has fine pointed leaves with toothed segments. In the eighteenth century double-flowered varieties were developed from it but the hybrid flowers we know as Chrysanthemums, *C. sinensis x indicum*, though grown in China for two millennia, did not reach the West till the nineteenth century.

Red Hot Poker is to our eyes so typical a flower of the cottage garden that it may come as a surprise to learn that all species of *Tritoma*, or *Kniphofia* as we must now call them, come from South Africa. The original Torch Lily was brought here as *Tritoma uvaria* from the Cape of Good Hope, then still under Dutch rule in 1707. Perhaps that is why the name of Tritoma has been changed in honour of Professor Kniphof (pronounced, as near as I can get, 'nip-hofe', not 'niff-off' to the embarrassment of any plain English speaker in a garden shop) of Erfurt, who produced a twelve-volume herbal illustrated by nature printing, an inked imprint of the pressed flower itself, in 1764.

To include them in this book is stretching a point because they were kept indoors as stove plants for over a century and only planted outside for the first time at Kew in 1848. However, proving hardy, they then enjoyed such popularity, and spread throughout Britain so that by the end of the century almost every garden, whether by castle or cottage, could boast of its bed of Red Hot Pokers. Their fall from favour due to a virus root rot has been as rapid as the firework they resemble, up like a rocket with a burst of popular acclaim, dying away in a few forgotten embers. 'Kniphofias? You mean Red Hot Pokers? Ugly old weed traps — such a waste of space and just dead leaves for most of the year. Granny thought the world of them but of course we threw them out years ago to make way for a new peat bank.' Every conservationist knows the gist of the story.

In many cottage gardens an old wall offers a dry habitat to many plants not otherwise accommodated in an ordinary patch of fertile soil, the prime object of a new gardener. As one looks at the many rainy days of an English summer, desert conditions occasioning the development of succulents do not come readily to mind yet those are what one gets on sand-dunes, rock or wall-head where water drains quickly away leaving plants exposed to rapidly rising sun temperatures and long periods of drought. Available water is conserved by thick leaves offering minimal evaporation surface in Xerophytes, as these succulent plants are known to science even if it's the sort of term ordinary folk can only remember for word games. Britain has several native stonecrops with five-petalled star-like flowers. Wall Pepper, *Sedum acre* L., is by far the commonest, spreading its little yellow cushions of blossom wherever conditions are sufficiently dry and its roots can gain a foothold. Wall, roof or roadside stone heap will all do as well as an intended rock garden.

Rock Stonecrop, *S. forsterianum* Smith, and Large Yellow Stonecrop, *S. reflexum* L., are also British natives though rare in the wild now and long cultivated in gardens. Rock Stonecrop has smaller, slender flowering shoots with leaves pointing up the stalk and short barren tufts, whereas Large Yellow Stonecrop has the lower leaves of its flowering shoots turned down and elongated barren rosettes of fleshy leaves which were formerly eaten in salads by the Dutch and recommended by Gerard.

White Stonecrop, *S. album* L., has reddish leaves and broad, flat-topped white flower heads. Though still found wild on a few sea cliffs it is the type of white stonecrop grown in gardens.

S. anglicum, Huds, English Stonecrop, known as Prick Madam in Tudor and Stuart times, is smaller, pinker and in spite of its name more usually to be found on acid rocks in Wales or the Scottish Highlands.

Pink Stonecrop, *S. villosum* L., came originally from wet rocks in the Pennines but is another species more commonly found now in gardens.

Roseroot and Orpine, *Rhodiola rosea* L. and *S. telephium* L., are our two larger native stonecrops. Roseroot, still common on sea cliffs in the West Highlands, has a flat tuft of yellow flowers but is more striking in its foliage, a stiff tier of overlapping leaves, grey-green and usually purple tinted. It was formerly prized for its root which, when dried, smells of roses and was used medicinally. Orpine is more widespread in damp woods throughout Britain, and in Fife where woods are fewer and damp, shady conditions less widespread, it manages to survive very well in roadside culverts. Its toothed succulent leaves are alternate as distinct to the opposite leaves of *S. spectabile* Bor., the more spectacular Ice Plant introduced from Japan in Victorian times.

Houseleeks, species of *Sempervivum* ('ever-living' in Latin), and especially *S. tectorum*, have been with us probably as long as stone buildings and roof tiles have been able to provide a durable habitat. There are many closely related species, some cobwebbed, all asking nothing more than a place in the sun and enough crumbling mortar to secure their roots. Planted above cottage doorways by the simple expedient of raising a tile and wedging in their rosette of fleshy leaves, Sempervivums have always been associated with good luck as a guard against fire, lightning and thunderbolts. Perhaps such associations account for Houseleek's old country name of 'Welcome home, husband, be you ne'er so drunk' and the thunderbolts could be domestic? Certainly if Jack fell down and broke his crown whether going up the hill for water or something stronger, the fleshy leaves of Houseleek, according to Margaret Brownlow, could be bruised and laid on burns and hurts to give relief.

Such old wives' tales so often have a basis of truth. As a painter rather than a biochemist I cannot evaluate the medical worth of the connection but I am indebted to my friend Lys de Bray for explanation of an annually beautiful phenomenon: 'The reddish colour that stonecrops and other succulents develop in a hot summer is caused by the presence of anthcyanine, which is formed by the sap to protect the delicate inner tissues from scorching sun.'

Cooler, shadier crevices on the wall tend to be colonized by Spleenworts and ferns such as *Ceterach officianarum*, Rusty Back Fern, common only in the South-West and a typical species of Wessex walls. It has leathery dark green fronds deeply lobed and covered in scales on the undersides. Silvery at first these turn rusty with age and with the round leaves of Pennywort, *Umbilicus rupestris* Dandy, such curious green plants form an instantly recognizable feature of the region, a truer part of its character than some garish blooms from an international mail order catalogue.

Ivy-leaved Toadflax, *Linaria vulgaris* Mill., or Mother of Thousands, has become another common plant of old walls through Britain, which is odd for such a seemingly frail little flower. It is believed to have originated in Southern Europe where it was used as a salad plant, perhaps like cress as a garnish because there's nothing much about it to eat. Gerard said he'd never seen it but, among Cyclamens, described: 'This plant hath greene cornered leaves like unto Ivie, long and small gaping flowers like the small Snapdragon', which fits. Thomas Goodyear is the first to refer to it definitely: 'I never saw this growing but in the garden of my faithful good friend Mr William Coys in Northokington in Essex and in my garden at Droxford of seeds receaved from him, anno 1618.' In 1633 it is called 'Bastard Italian Navelwort' in Johnson's edition of

Flowers of the dry wall-head: Thrift and Wall Pepper with Small Copper on it, Ivy-Leaved Toad Flax, Rusty Back Fern, Herb Robert and Houseleek.

Gerard and by 1640 Parkinson says it is going wild: 'it groweth naturally in diverse places of our land'. Its success in achieving such rapid colonization must in part be due to the curious ability of its seedhead to turn round. After flowering the shady side of the stalk expands more than the sunny side, thereby causing the ripening fruit to twist away from the sun and towards the wall. Thus one can see the fruits nosing like tadpoles to find lodgement in each crack and crevice rather than risk wasting their seed on the wind with only a small chance of survival.

Armeria maritima (Mill.) Willd., Thrift, was originally included as Sea Pink under *Caryophyllaceae*, and with Sea Lavender called Statice because it means 'stop' and they were supposed to stabilize shifting sand. With their papery bracts the species are botanically akin but Thrift is happiest where it can get its long root firmly anchored in a rock crevice, and develop its evergreen grasslike leaves almost impervious to sun, wind or salt sea spray on a peaty cushion of previous years' growth. In ordinary garden conditions it is liable to damp off and fall a prey to any ranker shoots pushing through its neat tufts.

This characteristic growth made Thrift an obvious choice for Tudor Knot gardens though, even to such a labour-intensive fashion of gardening, Thrift had its drawbacks. Thus Parkinson warns 'Yet these inconveniences doe accompany it; it will not onely in a small time overgrow the knot or trayle in many places, by growing so thicke and bushie that it will put out the forme of a knot in many places; but also much thereof will dye with the frosts and snowes of Winter, and with the drought of Summer, whereby many wide places will be seene in the knot, which doth much deforme it and must therefore be yearly refreshed: the thickness also and bushing thereof doth hide and shelter snayles and other small noysome wormes so plentifully, that Gilloflowers and other fine herbes and flowers being planted therein are much spoyled by them and cannot be helped without much industry and very great and daily attendance to destroy them.'

Thrift was obviously easier to grow in the cracks of paving in courtyard or on a terrace and in humbler gardens retains its popularity as an edging plant. No one can suggest any reason why it should have been called Thrift but the general frugality of its growth.

Evening Primrose, species of *Oenothera*, provides a good example of the quagmires of proper botanical identification and taxonomy by which the amateur may become engulfed over a seemingly innocent and well-known flower. Evening Primrose? Yes, of course, we know it in every cottage garden – big primrose-coloured flowers soft as poppies, opening after tea

Love-in-a-Mist and, (clockwise), Evening Primrose, Rose of Sharon and Orpine.

RB 9/82

and shrivelling in next morning's sun to be replaced by more up the sticky red stalk — biennials making a striking rosette of leaves in their first year — about the same size and with leaves rather similar to bolted spinach — likes growing on well-drained chalk, railway embankments, sand-dunes and that sort of place — where's the difficulty?

Fifteen species have been recorded as growing wild in Britain, let alone garden species which have not yet been known to escape. Some seem to be European natives, some come from North America, some from South, some in the eighteenth century, some in the nineteenth, some in the twentieth. The latest account of them and their hybrids, trying to impose scientific order where all seems fluid, covers thirty pages in *Watsonia*, the Botanical Journal. According to Doctor Rostanski of the Silesian University at Katowice in Poland, both the sizes of petals and stigma lobes varies by half in flowers taken from the same plant at different times of the year and the numbers of anthers, sepal tips and other vital statistics are even more variable. Where a genus is thus obviously still in a state of evolution, botanists tend to divide according to temperament between the 'splitters' and the 'lumpers'.

Although *Oenothera biennis* is the original European native species most commonly found today, the first to be brought into Britain were collected in Virginia by John Morus and figured by Parkinson in 1640. But, as Dr Rostanski points out, 'It is certainly not *O. biennis* as was thought by Linnaeus and most later authors.' He draws our attention to the 'transitional part between ovary and base of hypanthium', a detail insignificant to anyone but an expert. Once these shades of difference are pointed out one can see that these charming flowers which are liable to turn up in any dry corner as escapes from imported wool or shoddy in the vicinity of docks, railways or roadsides, may belong to a variety of evolving types. If the cottage garden habitat is to their liking, species of Evening Primrose such as *Oenothera cambrica* L. may settle. This particular species is noticeably redder in the stalk than most of the others, and has colonized sand-dunes at Berrow in Somerset.

Snapdragon, *Antirrhinum majus* L., though a native of Southern Europe, was well established here by the time the first plant lists were compiled by Tudor gardeners. Its natural preference for a warm, well-drained situation was observed by John Hill in 1757 when he noted that it 'will live on an old wall and propagate itself from year to year' whether by seedlings or on a perennial rootstock surviving for several years in the right site. This hardiness has been sacrificed by modern horticulturalists who prefer a dumpy mass of variegated colours to the taller more spare

Snapdragon flanked by Toadflax and purple Toadflax.

elegance of the old plant, usually crimson or, as in the form 'pictum', a cream tube with a crimson lip. Bee guides to this lip form a marmalade mouth for the 'dragon' which is so firmly shut that unless the tube is pinched sideways by childish fingers it will only open naturally for pollination to the weight and thrust of large bumblebees.

Our native Toadflax, *Linaria vulgaris* Mill., common still in summer on many a grassy bank, is yellow with a long spur and an orange lip. Alas for the name! The harmless retiring toad only too often came in for trouble in a superstitious age as a wicked dragon, and the thin leaves are very like those of flax. Though originally an introduction from Southern Europe before records began, Purple Toadflax, *L. purpurea* (L.) Mill., a most elegant growth with even its linear leaves taking on a purple sheen in a dry summer, has been accepted into our flora and thrives in any dry corner of a garden path or wall-head.

Centranthus ruber (L.) DC, Valerian, another native of Europe, has been long enough settled here also to have become accepted into our flora. Although its leaves were eaten in Italy as a salad plant and the seed was formerly used in embalming, Red Valerian seems never to have had any healing application. A white form was developed in 1604 but it was as a picturesque wall flower that Valerian was esteemed by the grotto builders of the next century.

Verbascum comprises about ninety species of Mulleins, as they are commonly known, half a dozen of which are British wildflowers ranging from the stately *Verbascum thapsus* L., Common Mullein, or Aaron's Rod, with primrose yellow flowers massed up its white woolly spire, to *V. blattaria* L., Moth Mullein, with solitary flowers dotted sparingly up its dainty stalk. As Parkinson put it graphically three hundred years ago, 'many goodly yellow flowers with some small purplish threads in the middle, the ends whereof are fashioned somewhat like as if a Flie were creeping up the Flower'. They hybridize freely to the despair of botanists attempting to decide classification by the colour of their filament hairs. In the cottage garden this need not be our prime concern as the large rosettes of grey foliage will place themselves with inspired abandon anywhere dry and limy, in cracks in paving, on wall-heads and on well drained banks so that it matters little which species comes up to flower in the end. Candlewick and Hag's Taper are country names which show the old use of *V. thapsus*, meaning wand, as a lampwick or torch dipped in tallow, *V. blattaria*, Latin for cockroach, was supposed to drive them away at the same time as the plant was attractive to moths. To animals all Mulleins are poisonous.

On dry walls in late summer: Red Valerian, Yellow Corydalis and a migrating Clouded Yellow on Mullein.

Fumitory species were originally 'Fumiterre', literally 'earthsmoke', a name variously explained according to Alice Coats 'as being due to an early belief that the plant sprang up spontaneously without seeds; the use of it as a fumigant, the smoke of which would exorcise evil spirits or because it makes one's eyes water like smoke'. We have about a dozen natural species with pinky-white flowers like a shoal of tiny fish up the stalk. Really weeds of the cornfield, they can be relied on to spring up on any dug ground in the garden and do little harm. The closely allied Yellow Fumitory, *Corydalis lutea* (L.) DC, grows wherever it can find a crack in a wall. Originally sub-Alpine, it doesn't mind damp and lack of sun, so it has had a welcome place in many a darkened and unconsidered corner of old gardens for a long time. Gerard knew it as Bunnikens Holwort. Parkinson grew it, and Miller called it 'a very proper plant to grow in rockwork or upon old walls or buildings'. The leaves are not unlike those of that Victorian favourite Maidenhair Fern, so once Corydalis can get under the staging of an old conservatory, the pair of them stay forever like visitors who don't know when to go.

HARVEST TIME

Several species have been introduced to cultivation through cloth making prior to being encouraged as garden plants.

Teasel, *Dipsacus fullonum* L., for example, is familiar as an indigenous wildflower of calcareous grassland to anyone driving out of London through the chalk escarpments of the South-East. Teasel's natural northward distribution peters out at the Yorkshire Wolds, but, because it was used by weavers for brushing up the nap on cloth, it has tended to survive round the little weaving towns of the North-East wherever the old lime mortar of derelict buildings makes a suitable habitat. Cupar, Fife, is the most northerly station cited by the *Botanical Atlas* but I have found it further north in Angus at Friockheim and Brechin. The former was a settlement of German Protestant weavers and the latter has a family bakery still selling 'heckles', a kind of shortbread originally pricked with a weaver's comb just as in the abstract sense one needles a politician. Heckling improves the cloth where wire brushes might tear it so Teasels are still used in weaving, though nowadays they are grown commercially only in Spain, our own harvest fetching higher prices with dried flower arrangers. You may see the results dyed hideous colours for sale on market stalls.

In addition to being one of the most spectacular plants, Teasels have the interest of being in part carnivorous, with bracts attached to the central stalk in such a way as to form reservoirs of rain water several inches deep at the base into which any creepy crawlies descending the stalk must fall to be ingested later.

Woad, *Isatis tinctoria* L., grows naturally on a few limestone cliffs in Wessex and is introduced elsewhere. It too makes a handsome garden plant with its great spreading candelabra covered first with a cloud of tiny yellow star flowers and secondly with seedheads hanging like row upon row of pea-green paper clips. There is no hint of the famous blue dye used as war paint by the Ancient Britons till the sap begins to drain away in the autumn and then, as the sere leaves begin to yellow, a violet tinge appears at the edge which becomes more apparent as the leaves rot. This natural process of woad is still used as a starter for commercial fermentation of indigo which, as a dye, has now superseded woad.

Dyer's Greenwood, *Genista tinctoria* L., is a wildflower, rather like a miniature Broom or Petty Whin of Scottish moors, which grows into a shrubby clump about two feet high covered in similar yellow pea flowers. Because it behaves itself, being neither invasive nor over-prodigal with its children, Dyer's Greenwood often finds a place as a background plant for rock gardens or in the front of semi-wild shrubberies with people who have no idea of its useful cottage garden past. The yellow flowers yielded

Weavers' plants: Teasel, Woad, Blue Flax and Dyer's Greenwood.

yellow dye and when combined with blue woad a green, hence its name.

Rubia peregrina L., Madder, is a native of the hedgerows and verdant sea cliffs of the South-West where this most robust of the cleavers scrambles about by means of its downward-pointing spines. *R. tinctorum* L., Dyer's Madder, is softer with its leaves net-veined on their underside and flowers brighter yellow rather than greenish yellow. It came originally from the Mediterranean where since antiquity it was grown for its roots, source of the celebrated dye Rose Madder. It survives wild in Central Europe as a relic of cultivation and sometimes as a garden escape from cottage industry in the milder parts of Britain.

Linum usitatissimum L. was the species of Cultivated Flax once a common crop in the cooler, wetter climate of the north but now almost extinct. Only in place names is it possible to trace the extent of this once thriving industry. Wakefield, Walkland, Waulkmill all refer to places where the flax crop was laid out exposed to the weather and 'walked', rolled and rotted till its fibres were ready for use. It is always annual, without non-flowering shoots and, according to Polunin, 'Its origin was unknown, occurring as a casual from cultivation almost throughout Europe. Flax produces linen, probably the oldest known textile, and it has been cultivated for this purpose since early Egyptian times. It is used to make cloth, ropes, sails, nets, etc., but is now largely superseded by cotton and the new man-made fibres. Linseed oil is obtained from the seeds; it is a drying oil used in varnish and printers' ink; the brushed residues form oil-cake which is an important food for cattle.' *L. anglicum* Mill. is the closely related perennial Blue Flax of East Anglian chalk grown more usually in gardens where for months on end it can be relied on to open a new crop of soft blue flowers every day.

The field was never far away from the old cottage garden. In fact the cottage itself was usually fitted into an odd corner of fertile farmland so that many of our garden flowers jumped the hedge and were made welcome from the fields provided they behaved themselves: Cornflowers and Chicory do, Poppies do not. Cornflowers, *Centaurea cyanus* L., sometimes called Bluebottle or Hurt-sickle because it was so common amongst the corn in Constable's day that its tough stalks blunted the reaping hooks, has been in decline ever since and is now listed as a threatened wildflower. It has been cultivated in gardens since Tudor times and pink, white and purple forms were developed by the seventeenth century. Its strawlike blue flowers were favourites in eighteenth-century France, where they were dried for winter decoration and as such enjoy a revival today.

Flowers of the cornfield, Red Poppies, Chicory, and Cornflowers among Barley.

C. montana is the larger perennial Cornflower of the herbaceous border. Introduced from the Pyrenees in 1595, it has a stronger-leaved growth which often survives as a garden escape.

Cichorium intybus L., Chicory, is a wildflower of the chalk down. Until a few years ago its pale-blue star-flowers were common at the fringe of the cornfield till improved farming made it increasingly rare in this country. As it is exactly the same species as the plant which is forced to provide long tender salad shoots in winter, it seems a pity that it is not grown more popularly in the modern cottage garden, contributing as it does to both kitchen and flower border. Red Poppies too were once the scourge of the farmer, due to their habit of coming up wherever the ground has been disturbed. There are two species, *Papaver rhoeas* L., with round seed capsules, and *P. dubium* L. with long ones. The seeds, as eaten on French bread, are capable of surviving dormant in the soil for years even though it be the churned-up mud of Flanders trenches. Thirty years ago when I worked as a harvest help, bagging-boy on the combine harvester, there were always in addition to two sacks filling with grain, a third one for 'lights'; odd seeds, green peas and bits for the hens and a fourth bag for poppy seeds which filled up every two or three circuits of an ordinarily clean field. Alas! Now that improved seed cleaning and selective sprays mean that there are no longer sufficient impurities to dirty the river of grain pouring into a modern combine's tank, even Poppies, the hardiest survivors, are feeling the strain and becoming less common.

Never again shall we see the Poppy-spangled fields that were taken for granted by painters. One thinks of Monet's impressionist dots of colour and Van Gogh's swirling vegetation that would spell ruin for any Common Market farmer in the fields of today and I should like to think that the time has come for the more perceptive private gardener to widen his tolerance and acceptance of such refugee species. Countless millions of Poppies and wild oats in a field are an agricultural problem but surely a splash of red in a wild garden (with the prospect of poppyseed rolls for breakfast?) or a few wild grasses arranged in a vase on a city shelf are to be judged by different standards as 1984 looms over us?

Papaver somniferum L., Opium Poppy, is another wildflower of cultivated land with an ancient history. According to the Greeks Somnus, God of Sleep, devised it for Ceres the Corn Goddess to stop her worrying about the harvest. It also signified fecundity with *c.* 32,000 seeds in each capsule according to Linnaeus. Although the showy if short-lived flowers in varying shades of pink and mauve were admired in Tudor times and the handsome glaucous foliage was a favourite with Dutch flower painters, Gerard, at least, knew all about opium addiction: 'It mitigateth all kinds

Opium Poppy and (clockwise), Wild Oats, Corncockle, Catmint, Brome Grass and Fumitory.

of paines but it leaveth behinde it often times a mischiefe woorse than the disease itself. Opium somewhat too plentifully eaten doth also bring death.'

By 1800, however, opium in Britain was a commercial crop. I quote Alice Coats: 'Mr Young calculated that an acre of poppies . . . would yield 56 lbs of opium and that poppy seeds being pressed yielded 375 pints of salad oil.' Opium at that time sold at 22 shillings a pound and it was estimated that 50,000 pounds were annually consumed in this country. In 1823 Miss Kent, writing about this Poppy, casually mentions in passing that 'the solution of opium in spirits of wine is now called laudanum, or loddy, so much used instead of tea by the poorer class of females in Manchester and other manufacturing towns'.

As one surveys the drug scene today one knows not whether to be sorry or glad at modern ignorance and idleness as, on the other hand, one looks at drifts of Poppies in cottttage garden and rubbish dump all un-harvested whether for salad oil or darker purposes. A nip of loddy sounds so cosy!

Corncockle, *Agrostemma githago* L., was one of the scourges of the medieval cornfield. 'Its seeds are poisonous due to the presence of saponins and if they contaminate flour in any quantity it may become dangerous to livestock and humans', says Polunin. Millers generally had a bad reputation due no doubt too often to the substitution of bad grain for good and Corncockle would have played its part.

In the nineteen-thirties, however, this formerly common wildflower was already rare. As a child in Suffolk I remember its spurred bright pink flowers surviving as casuals in the sort of odd dry corner where one found Henbane and Viper's Bugloss growing but, alas, its decline continued. The big black seeds were no match for modern grain-cleaning riddles and Corncockle became officially extinct during the seventies. My friend, the late Mrs Swanbourne, Botanical Recorder for Wiltshire, unwilling to accept this fate for such a handsome member of our flora, obtained seed from identical plants in Holland and began growing them in her garden. She gave some to me and I have given them to anyone who thought they might be able to grow them. Obviously not all will do well and farmers and bakers need have no fear that Corncockle will again become the menace it was in Chaucer's day but it may point the way forward for cottage gardeners in the twenty-first century. As factory farming becomes ever more efficient to feed a growing population, let us hope there will be other Mrs Swanbournes, redoubtable ladies in English villages, who in rescuing just one species make their small gardens significant as nature reserves.

Nepeta cataria L., Catmint, is very different from the various garden species and mauve hybrids passing under the general name of Catmint. *Nepeta cataria* is upright, not sprawling, with downy dull green foliage and small white, pink spotted flowers. It is true Apothecary's Catmint, strong scented and flavoured. Though I have lost trace of its author, I remember the story of the restaurateur who after a disastrous evening was just about to close in despair and give the *plat du jour* to his cat with a pinch of Catmint added for the sake of her imminent kittens when a stranger came in out of the rain wanting something to eat. As the hour was late and the dish was ready the stranger shared the cat's food. Of course he declared his impromptu dinner to be the most inspired cuisine and turned out to be the Michelin inspector. The restaurant won its coveted third star. Its *specialité de la maison* became famed throughout the region but as Monsieur and Minou kept their secret, no gourmet could ever quite place its subtle flavour.

Our native climbing peas have winged stems, both the rare Marsh Pea, *Lathyrus palustris* L., of the Somerset levels and *L. sylvestris*, the pink flowered species found on sea cliffs.

L. tuberosus L. with an angled stem has grown wild in Essex for so long that it is known as the Fyfield Pea. It is now generally conceded to be an early introduction from Holland where it was grown for the sake of its edible tuberous roots. This is not so odd when we remember that our native Sea Pea, *L. maritimus* Bigel., now known as *L. japonicus* Willd. because it is the same circumpolar species as that of the North Pacific, was eaten by the citizens of Aldeburgh to sustain themselves in famine and *L. montanus* Bernh., Bitter Vetch, was eaten as Cairmeal by Highlanders on the move to assuage hunger in the glens. To grow *L. tuberosus* for the ornament of its pink or rarely white flowers is a recent luxury.

L. latifolius L., Everlasting Pea, is another introduction much admired in old gardens. Gerard liked its 'most beautiful flowers like those of the Pease the middle part whereof is a bright red, tending to red-purple in graine, the outer leaves [petals] somewhat lighter inclining to a blush color'. Parkinson commended its 'pretty sent or smell'. In 1682 Nehemiah Grew discovered that the flowers steeped in spirits of wine produced a brilliant blue pigment similar to that of the fading flowers Bitter Vetch, but by 1807 MacDonald's *Complete Dictionary of Practical Gardening* declared that it was 'too large and rampant for the borders of the common flower garden' so it became demoted to shady corners and the cottage garden.

By then *L. odoratus* L., Sweet Pea, was able to take over. First recorded by Father Franciscus Cupani in *Hortus Catholicus* 1697 as a Sicilian wildflower, seeds of this newly discovered exotic were sent by Father Cupani in 1699 to Dr Robert Uvedale, Master of Enfield Grammar School, one of the earliest gardeners to possess a range of hot houses. This is described as 'Standard violet, wings and keel bluish white or pink' by Polunin. All commended the scent 'somewhat honey-like and a little tending to the orange flower smell' but as the flowers are self-fertile, chances of cross-pollination are few and little progress was made till the late nineteenth century when selective breeding by Henry Eckford produced the Sweet Pea as we know it.

Jasminum officinale L. is one of the most ancient of flowers. In the ninth century, the Chinese knew it as 'Yeh-hsi-ming' and that it was of foreign origin. In Arabic, it was 'ysmin' and was carried along the caravan routes of Asia. In 1825, Russian botanists claimed to have found it growing wild on the Caucasian shores of the Black Sea near the old silk road to

Samarkand. In Persia it was 'ja sémin', which try-syllabic form evolved further in Jerusalem to the Garden of Gethsemene. The plant's importance lay in its oil of 'Jasme'. Dioscorides had the white flowers steeped in sesame oil. A more efficient process still used today is effleurage whereby fresh flowers are daily patted into layers of fat, which when saturated is then dissolved in alcohol and distilled.

White Jasmin was commonly grown in London before 1548, and by 1629 Queen Henrietta Maria had introduced the tender Great Spanish Jasmine and Jessamy gloves were in fashion, Cavaliers paying a shilling an ounce for oil to scent their fingertips. Such vaingloriousness is a far cry from the Jasmine-bowered porch scenting the cottage on summer evenings. In contrast the Imperial Yellow Chinese Winter Jasmine discovered near Pekin and introduced by Robert Fortune in 1844 is scentless.

The cottage garden climber, par excellence, must surely be Honeysuckle, *Lonicera periclymenum* L. 'Periclymenum' means 'to roll round about' in reference to its growth habit and Adam Lonitzer or 'Lonicer' was a physician of Frankfurt who published a curious natural history of plants in 1555.

Among early references to it we find the Earl of Lincoln had Honeysuckles in his Holborn garden, now Lincoln's Inn, before 1286. The graceful silhouette of its clustered buds was a favourite decorative motif of the Greeks and was borrowed in turn by Robert Adam along with other motifs originally homely in their inspiration such as acanthus leaves, egg and dart mouldings, flower swags and chains of barley husks to grace the new classical elegance of his plaster-work ceilings. Honeysuckle flowers have evolved their trumpet shape, similar to that of many tropical flowers pollinated by humming birds, in partnership with their pollinators, the long-tongued night-flying Hawk Moths.

Growing wild in the wood, Honeysuckles can do much more damage than ivy, entwining small saplings so tightly that they take on a corkscrew shape. In a more settled age, countrymen used to train them up deliberately, selecting Gean or Ash while still sufficiently supple to bend the top over and tie it to form a crook handle, in the knowledge that in a few years they would find it grown into a barley sugar walking stick. Hence the name Woodbine goes back to the Anglo-Saxon 'Wudubinde' and was used indiscriminately for ivy and convolvulus just as in the Middle Ages 'hunnisuccles' could mean nectar-rich clover too. 'Woodbyne which beareth the Honeysuckle' is Murray's usage separating the plant and its flower.

Duke of Argyll's Tea Tree, Honeysuckle and Damsons with Red Admiral enjoying ripe fruit.

Honeysuckle arbours were already popular by Tudor times as mentioned by Shakespeare in *Much Ado about Nothing* 'Where honeysuckles ripened by the sun, forbid the sun to enter'. Other varieties such as the Early and Later Dutch (var. *belgica* and var. *serotina*) had been brought over by 1715. Because the paired leaves of *L. caprifolium*, Perfoliate Honeysuckle, are joined at the base and its flower clusters are stemless, this singularity was described by Gerard thus: 'the leaves and flowers do resemble sawcers filled with the flowers of woodbinde . . . three or fower saucers one above another filled with flowers as the first', which decreasing formation earned it in an age of sailing ships the name 'tops and top-gallants'.

L. etrusca, a similar South European species, with long stemmed flower clusters was introduced in 1750 and their natural hybrid *L. x italica* was taken to the American Colonies where it escaped and when 'discovered' by later botanists was mistakenly called *L. americana*.

Another Eastern introduction which has settled unobtrusively into the cottage hedgerow is *Lycium halimifolium* L., Boxthorn, or curiously named Duke of Argyll's Tea Tree. This perpetuates either a mistake of the confused notes and labels kind which besets even the most methodical gardeners or else a deliberate fraud by wily Chinese plantsmen who had no intention of sharing one of their most valued species with the West. The shrub which Collinson recorded in 1752 as sent from China to the Duke as the tea tree is in fact a Solanum, dingy flowered and useless if not downright poisonous. It is, however, one of life's survivors. 'Lycium' means 'thorny shrub' though Dioscorides used the name probably meaning Berberis. The misnamed Tea Tree is well naturalized on the sandy bluffs and shingle banks of the south-east coast and forms a good thickset or impenetrable hedge in places where nothing else will grow.

Because the flowers have a slight scent, eighteenth-century gardeners used it for bowers where its downward trailing shoots like a kind of thorny Winter Jasmine could be seen to best advantage. But toughness is its most valuable characteristic. Once established, it grows stubbornly resistant to rabbits, sheep or the worst that even the winter of 1981/2 could do to it. The most flourishing old plant I know grows at Kindrogan Field Centre, well up a Highland glen on the back of a barn and I know of several still surviving, disregarded, in cottage garden hedgerows long after the cottage itself has become a ruined shell.

Mankind seems usually to have got it wrong over daisies. *Compositae* with their deceptively simple rayed florets, defeat most botanists because they are too successful, hybridizing with a free abandon that leaves scientific nomenclature puffing along miles behind when Asters alone comprise over three hundred species. Early botanists, recognizing the attractive starlike quality of these showers of late summer daisies, called them Starworts or Asters but now the name is reserved for dumpy annuals, *Callistephus* species or China Asters introduced by mid-eighteenth century horticulturalists, which they have now bred to look like pink and purple ostrich feathers of poor quality. So much for Starworts, which came to us first in Tudor times from Italy and which can still be seen on any Italian roadside as a cloud of white stars above leaves which turn attractively crimson with the first frosts. We have our own indigenous Sea Starwort, *Aster tripolicum*, a mauve daisy still common on saltings and once cultivated according to Gerard as Serapia's Turbith although it had no particular herbal use.

Most of our daisies, mauve or pink stand-bys of the autumn garden, came from the American colonies in the eighteenth century. 1752 marked the introduction of the revised Gregorian Calendar making allowance for a nicer adjustment of leap years. Though the mob rioted, chanting 'Give us back our eleven days' and May Day has been cold, showery and unpoetic ever since, the corresponding seasonal adjustment in autumn of Michaelmas meant that it coincided with the flowering of the new mauve species of Asters coming in from the New World, henceforth known as Michaelmas Daisies. Again luck was against the botanists. In 1687 a German, one Herman, had called them very properly *Aster novi-belgiae* after the Dutch settlement in America where he found them but, alas, its little town of Niew Amsterdam became New York which curtailed the suitability of that name too.

A. novi-angliae, a pink species, came from New England in 1710, and *A. grandiflorus*, very late flowering, in 1720, but as a family they had to wait for a proper appreciation until William Robinson and the Edwardian heyday of the herbaceous border massed their starry clouds to good effect and made the development of modern cultivars worthwhile. Today, when the stock of such labour-intensive gardening has usually been curtailed in favour of shrub borders, the older forms and species are usually to be found on neglected banks, old rubbish dumps and railway cuttings where their more gentle charm lures the last butterflies of the season.

Like Shasta Daisies ousting Marguerites, tougher more splendid American species of Golden Rod, *Solidago canadensis* L. and *S. gigantea*

Ait., have replaced our humble European sort in all respectable gardens.

Our own original Golden Rod, *Solidago virgaurea* L., derives its scientific name from its medicinal properties, 'solitare' is Latin for to heal, but Gerard amusingly charts its decline as a healing herb: 'I have known the dry herb which came from beyond the sea sold in Bucklersbury [the herb market] in London for halfe a crowne an ounce. But since it was found in Hampstead wood even as it were at our townes end, no man will give halfe a crowne for a hundredweight of it: which plainly setteth forth our inconstancie and sudden mutabilitie, esteeming no longer any thing, how pretious so ever it be, than whiles it is strange and rare. This verifieth our English proverb, "Far fetched and dear bought is best for Ladies".'

Human nature changes little though nowadays would any men be so chauvinist as to exclude their own sex from Gerard's strictures? Meanwhile our European Golden Rods have been augmented by several stronger flowering species from the American colonies but they all have a similar dense network of surface roots which impoverish the soil and justify their relegation from the flower border to the rubbish dump. Golden Rod, however, is not so easily to be put down and unless the roots are burned, golden cascades of blossom to delight insect life in autumn mark the site of many an old garden tip.

Hypericums, producing their sun-like flowers at the summer solstice, have always been well-regarded as a deterrent to black magic long before Christianity took over this beneficient aspect and called the plant St John's Wort. Because the leaves of *Hypericum perforatum* L. are covered with pellucid red spots, in reality tiny oil glands, folklore held that they had been pricked by the Devil out of jealousy for the plant's healing properties. A blood-like red oil can be cooked up from the sap so the medieval Doctrine of Signatures suggested the plant as a cure for deep wounds. Gerard gives a recipe and I, for one, am constantly glad to be saved by a mere accident of time from such ministrations of old apothecaries. *H. androsaemum* L. is popularly named Tutsan, a corruption of 'Toutsan' or 'health all round', and is a British native common in West Country woodlands. On the other hand the large-flowered *H. calycinum* L., Aaron's Beard, or Rose of Sharon, known to Gilbert White in 1790 as Sir George Wheler's Tutsan, was found by the latter near Constantinople in 1675 and sent to Robert Morison at the Oxford Botanic Garden whence it soon became a popular ground cover for landscape gardeners and known as 'Park Leaves'. Today, because of its accommoding disposition it has descended to being a favourite plant for traffic islands. *H. patulum* from

Canterbury Bells and (clockwise), Golden Rod,
Michaelmas Daisies, the fruit of Bladder Cherry
known as Chinese Lanterns, and Marigolds.

China in 1862 being not stoloniferous, is the parent of our modern hybrids.

Winter Cherry, Chinese Lanterns, *Physalis alkekengi* L., derives its scientific name from classical antiquity. 'Physalis' is Latin for a bladder but 'Alkekengi' was used by Dioscorides in the first century AD and the word is believed to derive from ancient Arabic roots. As Alken Kengy it was already familiar in the fifteenth-century English gardens, when the cherry-like fruit inside the lantern was once recommended for gout. For the last four centuries it is as a winter decoration that Physalis has been cherished. Parkinson, however, noted the plant's tendency to 'runne undergrounde and abide well enough' so it is out of favour in an age which prefers tiny gardens and plastic decorations indoors.

Tradescantia virginiana, Spiderwort, was so called because it was believed erroneously to cure the bite of a spider. It was one of the first plants to come to us from the English colony at Virginia but that too was in mistake as John Tradescant the Elder received it of a friend as Silk Grass. His son, however, put things right by making a voyage to collect plants in 1637 bringing back among other things Virginia Creeper. The original plant was blue but red, pink, white and purple varieties had been developed by the mid-seventeenth century and, given mild weather, Spiderworts will go on flowering till late into autumn.

Even before Linnaeus's time, the Saffron Crocus, *C. sativus*, had become sterile, to be propagated by root division. In our cold climate, unless frequently lifted and divided, the bulbs soon cease to flower, producing only grey-green ciliate leaves. Nowhere in the world is it to be found growing in the wild and Bowles considered it a possibility that *C. sativus* derived originally from the wild Greek variety *cartwrightianus*. In constrast, *C. speciosus* introduced by Bieberstein from the Black Sea Coast in 1800 is so hardy that it spreads freely from up to a dozen cormelets annually on each bulb. Each flower has stigmas almost as large and lax as those of the Saffron Crocus and lilac petals feathered in blue. Even in Scotland they are produced in such profusion that when I first saw them at Dalgairn in Fife when we came to the house in 1964 I thought it was pools of water lying on the grass below the crimson leaves of azalea bushes.

Though flowering concurrently and superficially similar, Autumn Crocus with their straplike leaves, belong with Irises to *Iridaceae*, and Colchicums with their lush green foliage to *Liliaceae*. Colchicums have six stamens to Crocus' three and three styles to Crocus' one, only divided at

Spiderwort or Tradescantia an early American introduction.

the top into three. Moreover Colchicums keep their ovary below ground so that the flower is carried on a long perianth tube rather than a stalk. The bulbs are so large and all-sufficing that they can produce a succession of blooms without any root growth to supply further nutriment in the flowering season. In the wild, seeds are distributed by ants attracted by the sugary outer layer but deterred from further damage to the embryo and cotyledon by colchicine, a powerful poison. Even in the gardens of the North during the last sunny days of autumn you can watch little Syrphus flies hovering in the concentrated warmth of sun reflected by the globe shape of the fully open flowers.

As the medicinal and culinary use of saffron has declined, so, according to Bowles, the importance of colchicine has increased, 'being used by cytologists to induce an increase in the number of chromosomes'.

Our own *Colchicum autumnale*, Meadow Saffron, sometimes called Pale Maidens in its usual rosy form or Naked Boys in its white, has spread from its native dry grasslands of Wessex to gardens all over Britain. Parkinson describes nineteen kinds in his *Paradisus* in 1629, but it is thought extremely unlikely that he knew so many personally and mostly copied them from Clusius's *Theatrum Florae*. The larger more splendid *C. speciosum* was described by C. Steven and introduced from the Caucasus in 1828. It is one of several species often tending to a chequered pattern known as tesselation when its flowers first open but as the colour of the style changes as well as the petals and the leaves have died off by midsummer anyway, the many closely related species of Colchicum are notoriously difficult to identify with certainty.

The cultivation of Colchicums is correspondingly easy: do nothing. Being wood lilies they resent disturbance but bloom with happy dependability in the sheltered corners of old gardens whose owners can resist the urge to tidy up their lush summer foliage and let well alone till the flowers burst forth from drifts of dead leaves like a final firework display in autumn.

In the shelter of a mixed hedge is Montbretia,
Rowan berries, Colchicum and Autumn Crocus.

As the days shorten and the gales of the autumnal equinox blow across the garden, hedges again show their worth. Whilst in grander places hedges of close clipped Yew may be centuries old, in the cottage garden, the hedge is usually a mixed one with Hawthorn, *Crataegus* species, predominating over lingering dry Beech leaves or Oak, Ivy, Field Maple, Ash, in a mosaic that is one of the quiet delights of the winter countryside. Many flowers linger a little longer on the sunny side of the hedge but Montbretia with its green sword leaves and flame-coloured flowers is especially typical of the season. Montbretia, named in honour of Antoine François Ernest Conquebert de Montbret, botanist to Napoleon's expedition to Egypt, is a corm of the genus *Tritonia*, of South African origin, one of which has now been accepted into the British flora along with Hottentot Figs, Buddleia, Himalayan Balsam, Pineapple Weed and various other aliens which, however much botanical purists may huff and puff, are clearly here to stay.

In superior gardens, Earlham hybrids, in growth more upstanding, their flowers a deeper orange and more numerous, are usually cultivated while the original species, *Crocosmia aurea*, is usually thrown away. However it isn't South African for nothing and the corms can survive long periods dry and rootless on the rubbish dump just as if it were summer drought on the veldt. Their delicacy reminds me of Freesias and I prefer the softer marmalade colour of just two or three flowers to the cultivar's harsh copper blooms slotted in by the dozen up a stalk rigid as a fishbone. As they are not completely hardy, Montbretias benefit from the shelter of a hedge – most gardens have one to screen the rubbish tip – and as they like drought conditions it is one of the few species to grow under soil-impoverishing Privet. It also thrives in the cuttings of the older parts of the London Underground system where one may often see its little orange trumpets and flag-like leaves brightening the smoky gloom of autumn bonfires along the line.

Among fruit trees Damson, included botanically with Plums under *Prunus domestica*, is perhaps the most typical fruit tree of the cottage garden. Apples and Pears have proved capable of constant refinement at the hands of the professional grower. What with grafting, pruning and training into cordon, espalier or fan they have now moved right up in the world, whereas unimproved Damson remains close to the Wild Plum or Bullace, *P. instititia*, sturdily on its own roots and suckering freely along the hedgerow. Unpruned and untended with its seasonal shower of small bitter black fruits of little use to anyone who isn't prepared to get down to real cooking; pies and plum puddings and the old country chores of

Christmas Roses, Holly, Mistletoe. Box hedges provide a sheltered source of food for Robin in winter.

jam and jelly making. Damsons don't lend themselves to modern marketing techniques; the bitter little cooking plums are going to make no advertising agent's fortune so why bother in a world which will accept big sweet plums and red apples tasting of cotton wool? Only a handful of people who think for themselves and won't put up with second best like the tartness of Damsons and see the point of fiddling about with all those little stones so let them continue. To use Damsons in the kitchen you have either to cut out the stones or to stew them and sieve the pulp in order to make purée or the inimitable Damson cheese or, if making jam, to have biddable child labour to extract the kernels from the pips. I remember sitting for hours of what seemed the endless days of childhood's summer with a hammer on the steps learning how not to hit my fingers in order to produce half a cupful of nuts which were then blanched and peeled for adding to Damson jam, well worth the time for the suggestion of almond flavour and for the excitement of finding slivers of ivory nut among the purple fruit. So because they have retained the natural resilience of wild species, Damsons have survived in old cottage gardens and suckers I have chopped off our own ancient trees and given away to friends are already bearing fruit.

Sorbus aucuparia L., Mountain Ash or Rowan, is a native species of the moorlands of Northern Britain. Perhaps because it is so often to be found clinging to a bit of rocky outcrop above a burn in the heather where nothing else will grow, it has been long held to be a talisman against witchcraft and encouraged to grow on either side of the garden gate. Certainly Rowan jelly is white magic when clusters of seemingly intractable little berries and a few windfallen apples are transformed into tart rose coloured jelly fit for the most discerning gourmet not to mention the pleasure of seeing panicles of creamy flowers all over the little tree in early summer.

The Holly and the Ivy of the old carol, *Ilex aquifolium* L. and *Hedera helix* L. respectively, were the original evergreens of our primeval woodland. Scots Pine, *Pinus sylvestris* L. came from the forests of the acid Highlands. Strawberry Tree, *Arbutus unedo* L., is allowable as a native part of the Lusitanian flora of south-west Ireland, otherwise the list is short; Yew, *Taxus baccata* L., Box, *Buxus sempervirens* L., and Juniper, *Juniperus communis* L. All the rest of our winter evergreens are comparatively recent introductions.

Holly and Ivy are interesting in that they produce different leaf forms on the same plant. The familiar prickly leaf of the Holly bush gives way to a smooth leaf once the tree is firmly established at a height no longer needing prickles to fend off browsing ruminants. Ivy, too, when it has climbed safely to the top of its host, develops woody branches with smooth leaves in place of the familiar five-pointed palmate type. Both were typical of the cottage garden hedge, giving it an interest in winter that a single species hedge can never have. Incidentally, though the roots of Ivy will eventually penetrate and pull down a building, it is the opinion of modern forestry that it cannot harm a healthy established tree: only when the host tree has died from other causes will the weight of mantling Ivy help its fall.

Juniper grows as a spire shaped shrub both on chalk downland and in old Caledonian pinewoods. Sometimes one finds bushes paired either side of a cottage garden gate or at the top of steps but as it doesn't lend itself so well to clipping and tends to burst open at the bottom when grown as a column, Juniper has never been so popular as Holly, Box or Yew.

Box was originally a common native of the downlands as at Boxhill in Surrey. Nowadays, it is usually just the name which survives, a ghostly reminder of woodlands we have lost; Box in Wiltshire; Boxgrove, Sussex; Boxley, Kent and Boxford and Boxted in Suffolk. There must be hundreds of other lesser known place names all confirming the erstwhile prevalence of Box. Now the Botanical Atlas lists only seven sites. In the wild state it grows to be a small tree, with fine-grained wood much in demand for drawer handles and inlay. In the garden its small closely set leaves and slow growth make it the perfect low hedging plant for any warm, well-drained site. It was planted originally for the utilitarian purpose of providing shelter to bring on spring vegetables in days before glass cloches. No longer needed for this purpose, the warmer microclimate under the shelter of the Box hedge is appreciated chiefly by birds who know that in hard weather, it is the best place to look for food.

Mistletoe, *Viscum album* L., is now, I suspect, known to a thousand urban dwellers for every one who still cultivates it in a garden. Formerly this shrubby parasite was believed to increase the yield of apple trees and I remember being shown as a boy how to slip the ripe squashy berries under cracks in the bark 'as if you're a bird cleaning its beak'. The Apple orchards of the West Country are the stronghold of Mistletoe but it will grow naturally on Poplar, Hawthorn, and other host trees. Because of its odd parasitic habit, Mistletoe never acquired Christian respectability like Holly but remained pagan, darkly suggestive of illicit goings on. As electronic man buys a sprig of Mistletoe for larks at the office party, we may smile at the reflection that in doing so he unconsciously performs a prehistoric fertility rite, with the idea of life in death brought down to us by Mistletoe fruiting as if by magic on winter apple trees in the cottage garden.

SELECT BIBLIOGRAPHY

Botanical Atlas of the British Isles Botanical Society for the British Isles (1976)

Botanists Garden, A John Raven (Collins, 1971)

Collecting Antique Plants Roy Genders (Pelham, 1971)

Collins Pocket Guide to Wild Flowers David McClintock & R. S. R. Fitter (Collins, 1955)

Complete Herbal Nicholas Culpeper (W. Foulsham, 1969)

Concise Encyclopaedia of Gastronomy Simon André (Collins, 1952)

Cottage Garden, The Roy Genders (Pelham, 1969)

Cottage Garden, The Anne Scott James (Allen Lane, 1981)

Cottage Garden and the Old Fashioned Flowers, The Roy Genders (Pelham, 1983)

Dictionary of National Biography (Oxford University Press)

Field Guide to the Insects of Britain & Europe Michael Chinery (Collins, 1973)

Finding Wild Flowers R. S. R. Fitter (Collins, 1971)

Flowers and their Histories Alice M. Coats (A. & C. Black, 1968)

Flowers of Europe: A Field Guide Dr Oleg Polunin (Oxford University Press, 1969)

Flowers of S.W. Europe: A Field Guide Dr Oleg Polunin & B. E. Smythies (Oxford University Press, 1973)

Flowers of the Balkans Dr Oleg Polunin (Oxford University Press, 1980)

Garden Book of Sir Thomas Hanmer, The Sir Thomas Hanmer (Gerald Howe, 1659)

Herball or General Historie of Plantes John Gerard (John Norton, 1597)

Herbs and the Fragrant Garden Margaret Brownlow (Darton L. & T., 1978)

Medieval Gardens John Harvey (Batsford, 1981)

New Concise British Flora, The W. Keble Martin (Ebury Press & Michael Joseph, 1982)

Old Shrubs & Roses Graham Stuart Thomas (Phoenix House, 1963)

Origins of Garden Plants, The John Fisher (Constable, 1982)

Paradisi in sole Paradisus Terrestris John Parkinson, (1629, Methuen reprint 1904)

R.H.S. Dictionary of Gardening, Vol. 4 Fred H. Chittendon (Ed) (Oxford University Press, 1951)

Wild Garden, The Lys de Bray (Weidenfeld & Nicholson, 1978)

Wild Garden, The (4th edition) Wm. Robson (John Murray, 1894)

APPENDIX

WILD ROSE CHARACTERISTICS

Dog Rose
R. canina

Prickles: curved
Leaves: smooth
Flowers: pink or white
Hips: red, hairless, egg-shaped

Common in South

Downy Rose
R. villosa

Prickles: straight
Leaves: downy
Flowers: deep pink
Hips: bristly, rounded

Commoner in North

Sweet Briar
R. rubiginosa

Prickles: curved
Leaves: downy
Flowers: deep pink
Hips: red, hairless, egg-shaped

recognized by its
sweet smelling
leaves

Field Rose
R. arvensis

Prickles: curved
Leaves: smooth
Flowers: always white,
         larger than Dog's
Hips: smaller than Dog's

recognized by styles
protruding above
yellow stamens

Burnet Rose
R. pimpinellifolia

Prickles: densely spiny
Leaves: small and rounded
Flowers: creamy white
Hips: purple – black

creeping in large low
patches by the sea in
the North